AF859707

Developing an Advanced Windows® Phone 7.5 App that Connects to the Cloud

Developing an Advanced Windows® Phone 7.5 App that Connects to the Cloud

David Britch
Francis Cheung
Adam Kinney
Rohit Sharma

978-1-62114-014-6

Contents

Foreword

The release of Windows® Phone 7, and the Windows Phone 7.0 SDK, provided great opportunities for building highly interactive and immersive applications. However, with the release of Windows Phone 7.5, and the Windows Phone 7.1 SDK, it is now possible to build many classes of applications that it was not previously possible to build. This release expands upon the capabilities of the Windows Phone platform by including many new features such as multitasking, local database support, Live Tile enhancements, deep linking into applications from notifications and Live Tiles, and an encrypted credentials store, to name but a few. These features, and many more, enable the building of even richer applications.

This guide will show you how to design and implement a compelling end-to-end application using the Windows Phone 7.1 SDK. Testability is a major focus of the guide, since as Windows Phone applications grow in complexity and size they can become difficult to test and maintain. The Model-View-ViewModel (MVVM) pattern provides a clean separation of concerns that not only makes applications easier to test and maintain, but also provides code reuse opportunities, and enables the developer-designer workflow. Windows Phone applications implemented using the Microsoft® Silverlight® browser plug-in are naturally suited to the MVVM pattern, which takes advantage of some of the specific capabilities of Silverlight, such as data binding, commands, and behaviors. The application presented in this guide combines a number of patterns in order to increase the testability of the application.

The guide also highlights a number of essential tools that can greatly increase developer productivity when building an advanced application. The Silverlight Unit Test Framework for Windows Phone and Silverlight 4 enables you to run unit tests on both the phone emulator and on real devices. The Silverlight for Windows Phone Toolkit contains extra controls that enable you to create even better applications using the Windows Phone 7.1 SDK. In the toolkit you'll find user interface controls like those found throughout Windows Phone, with components like toggle switches, page transitions, picker controls and more. Finally, the Prism Library for Windows Phone simplifies tasks such as binding commands to interface objects, linking methods to application bar buttons, notifying changes to object properties, and detecting changes to text-based controls in the view.

After reading this guide, you should know how to build your own advanced, loosely coupled, testable application using the Windows Phone 7.1 SDK.

Happy Windows Phone coding!

Sincerely,
Jeff Wilcox
Senior Software Development Lead, Windows Phone Team

For the majority of people in the world, their mobile phone will be their first computer. Here at Microsoft, we have recognized this for a long time. We entered the mobile market back in the 1990s and produced a reasonable offering, Windows® Mobile, which provided developers with a rich and compelling platform on which they could build a multitude of solutions.

In 2007 we recognized that our offering was no longer competitive and that to remain relevant we would have to reboot Windows Mobile in a drastic way. Our prime directive in this reboot was to deliver a fresh, compelling end-user experience targeted at making it extremely easy for people to perform the types of tasks we know people want to perform on their phones.

While we knew that we needed a drastic reboot of our focus and our end-user story on the phone, we didn't want to discard the things we knew we did well—specifically, our developer platform. Microsoft is a developer-focused company with many decades of experience delivering the required platforms and tools to developers. In building the application platform for Windows Phone we took advantage of the best developer tools and platform components that Microsoft had to offer. The result was a platform that's easy to use, and which enables developers to deliver compelling applications and game experiences that naturally extend the experiences of the phone itself.

However, tools and platforms alone are often not enough. For quite some time, the patterns & practices team has focused on providing detailed guidance about how to best use our tools and platforms. The first edition of this guide did an excellent job of teaching developers how to apply their skills to build a comprehensive end-to-end solution for Windows Phone.

Windows Phone 7.5 represents a very large expansion of the Windows Phone development platform. Therefore, we knew that a corresponding update to the patterns & practices guide was in order. We think this guide will provide you with everything you need to know to develop a wide range of compelling solutions that use the Windows Phone platform.

We'd also like your feedback. Let us know what you think of the Windows Phone platform at *http://wpdev.uservoice.com.*

Sincerely,
Larry Lieberman
Senior Product Manager, Windows Phone Application Platform

Acknowledgements

When I joined Microsoft® patterns & practices in May 2004, my projects were all related to client development: smart clients and web clients, mostly. At that time, we considered it natural to extend our guidance to mobile clients. The result of that was the Mobile Client Software Factory, which was released in July 2006.

As I was preparing for this project, I looked back at the work we did at that time, and I was surprised in two very different and opposite ways. First, the list of technical challenges to cover was surprisingly similar. Both mentioned things like UI design and dealing with networks. Second, modern devices are light years ahead of what we had at that time: much more memory is available, graphics processor units (GPUs) now exist, there are more sophisticated sensors and, of course, the cloud is a much more powerful back end. A lot has remained the same, and a lot has changed.

This book covers two extremes of the Microsoft Windows® platforms: the massive computing resources of Windows Azure™ and the personal, tailored experience of Windows Phone 7. As we were developing this content, I was reminded of the richness of the Microsoft platform, and the opportunities it offers to developers today. Ideas that were merely seeds in our imagination a decade ago or that were available to only large corporations with huge resources, are now accessible to everyone with a PC. I feel privileged to have contributed, even a little bit, toward making this happen.

This guide follows the same scenario-based approach we used in our previous three guides on Windows Azure development and claims-based identity. We created a fictitious, yet realistic sample that is used as a case study throughout the chapters. The sample and the guide are complementary. You will find that the guide covers tradeoffs and design considerations that go beyond what is implemented in code. Often, there are many ways to solve one particular technical challenge. We tried to surface those tradeoffs and the thinking behind our decisions to equip you with the tools to make your own decisions in your own environments.

In the code, you will find that we have chosen to solve many problems in ways that are new and perhaps unexpected. An example of this is the extensive use of the Reactive Extensions for .NET Framework for all the asynchronous network calls. We chose to do this because it is our mission to empower you with better tools and frameworks.

I want to start by thanking the following subject matter experts and main contributors to this guide: Dominic Betts, Federico Boerr, Bob Brumfield, Jose Gallardo Salazar, Scott Densmore, and Alex Homer. Dominic is a veteran of many patterns & practices guides. As I wrote before, Dominic has this unique ability to explain complex topics in simple terms without losing rigor. Federico has been a member of our team since the very first guide we wrote for Windows Azure and has both the technical expertise and the gift of empathy, an essential attribute required to write guidance. Bob is an outstanding developer who brought with him an incredible wealth of experience and knowledge about Microsoft Silverlight® browser plug-in development, the main framework used throughout the guide to build applications on the phone. Jose was one of the original developers of the Mobile Client Software Factory, and is a very experienced mobile developer who understands what developing guidance is all about. I feel very privileged to have worked with Scott every day—his knowledge spans an amazing spectrum, from devices to Windows Azure™ technology platform, which is exactly what we needed for this guide. For this project, he also brought the unique perspective of an iPhone developer.

I share two passions with Alex Homer: software and railways. We were very lucky to count on Alex's experience as a technical author; he contributed to the solid structure and flow of this guide.

Many thanks also to the project's development and test teams for providing a good balance of technically sound, focused code: Federico Boerr (Southworks SRL), Bob Brumfield (Microsoft Corporation), Scott Densmore (Microsoft Corporation), Chris Keyser (Microsoft Corporation), Jose Gallardo Salazar (Clarius Consulting), Masashi Narumoto (Microsoft Corporation), Lavanya Selvaraj (Infosys Technologies Ltd.), Mani Krishnaswami (Infosys Technologies Ltd.), and Ravindra Varman (Infosys Technologies Ltd.).

The written content in this guide is the result of our great technical writing and editing team. I want to thank Dominic Betts (Content Master Ltd.), Tina Burden (TinaTech Inc.), RoAnn Corbisier (Microsoft Corporation), Alex Homer (Microsoft Corporation), and Nancy Michell (Content Master Ltd.).

The visual design concept used for this guide was originally developed by Roberta Leibovitz and Colin Campbell (Modeled Computation LLC) for *A Guide to Claims-Based Identity and Access Control*. Based on the excellent responses we received, we decided to reuse this design in our most recent titles, including this one. The book design was created by John Hubbard (eson). The cartoon faces were drawn by the award-winning Seattle-based cartoonist Ellen Forney. The technical illustrations were adapted from my Tablet PC mockups by Katie Niemer (Modeled Computation LLC).

This guide, just like all our guidance content, was broadly reviewed, commented on, scrutinized, and criticized by a large number of customers, partners, and colleagues. Once again, we were extremely fortunate to tap into the collective intellectual power of a very diverse and skillful group of readers.

I also want to thank all of the people who volunteered their time and expertise on our early content and drafts. Among them, I want to mention the exceptional contributions of Shy Cohen, Istvan Cseri, Markus Eilers, Jonas Follesø, David Golds, David Hill, Yochay Kiriaty, Joel Liefke, Steve Marx, Erik Meijer, Miguel Angel Ramos Barroso, Jaime Rodriguez, Soumitra Sengupta, Ben Schierman, Erwin van der Valk, and Matias Woloski. A very special thanks is in order for the entire patterns & practices Prism team: Larry Brader (Microsoft Corporation), Bob Brumfield (Microsoft Corporation), Geoff Cox (Southworks SRL), Nelly Delgado (Microsoft Corporation), David Hill (Microsoft Corporation), Meenakshi Krishnamoorthi (Infosys Technologies Ltd.), Brian Noyes (iDesign), Diego Poza (Southworks SRL), Michael Puleio (Microsoft Corporation), Karl Schifflett (Microsoft Corporation), Fernando Simonazzi (Clarius Consulting), Rathi Velusamy (Infosys Technologies Ltd.), and Blaine Wastell (Microsoft Corporation).

Last but not least, I'd like to thank Charlie Kindel, the executive sponsor for this project.

I hope you find this guide useful!

Eugenio Pace
Senior Program Manager – *patterns & practices*
Microsoft Corporation
Redmond, WA, October 2010

Acknowledgements on This 2nd Edition

Here I am, almost exactly one year later, working on updating and improving this guide for building great Windows Phone applications.

The new Windows Phone includes some really great capabilities that we believe will help you develop much more powerful applications: a relational database, secure storage, and background agents, among many others.

The spirit of this guide has not changed, but if you read the previous release you will notice two very obvious differences: the title and the number of pages. We decided to change the title to better reflect the scope and focus of the content. This book contains guidance for building *advanced* apps. If your application is really simple, you might not need all the abstractions we have included here. But if high-quality software is your goal, this is definitely for you.

The guide is much slimmer now because we decided to remove all the introductory and general content on the Windows Phone platform and its capabilities. All that content was fine a year ago when product documentation was still being written and not widely available, but today there's plenty of content covering all that on MSDN. There is now excellent content available from both Microsoft and the extended developer community.

The result is a more focused guide that explores the design considerations of a relatively complex Windows Phone application interacting with a Windows Azure back end.

As before, we've reached out to the experts in the community to help us review, prioritize, and refine the content. In addition to the original advisors, I'd like to thank Amrita Bhandari, David Britch, Bob Brumfield, Francis Cheung, Scott Densmore, Jonas Follesø, Alex Golesh, Adam Kinney, Jesse Liberty, Rohit Sharma, Karl Shifflett, and Shawn Wildermuth. A very special thanks for Larry Lieberman and Jeff Wilcox from the Windows Phone team for their continued support.

Eugenio Pace
Principal Program Manager Lead – *patterns & practices*
Microsoft Corporation
Redmond, WA, November 2011

Preface

Windows® Phone provides an exciting opportunity for companies and developers to build applications that travel with users, are interactive and attractive, and are available whenever and wherever users want to work with them. The latest release of Windows Phone furthers this opportunity by enabling you to build many classes of applications that were not previously possible.

By combining Windows Phone applications with on-premises services and applications, or remote services and applications that run in the cloud (such as those using the Windows Azure™ technology platform), developers can create highly scalable, reliable, and powerful applications that extend the functionality beyond the traditional desktop or laptop, and into a truly portable and much more accessible environment.

This book describes a scenario concerning a fictitious company named Tailspin that has decided to embrace Windows Phone as a client device for their existing cloud-based application. Their Windows Azure-based application named Surveys is described in detail in a previous book in this series, *Developing Applications for the Cloud on the Microsoft Windows Azure Platform 2nd Edition*. For more information about that book, see the page by the same name on MSDN®.

In addition to describing the client application, its integration with the remote services, and the decisions made during its design and implementation, this book discusses related factors, such as the design patterns used, and the ways that the application could be extended or modified for other scenarios.

The result is that, after reading this book, you will be familiar with how to design and implement advanced applications for Windows Phone that take advantage of remote services to obtain and upload data while providing a great user experience on the device.

Who This Book Is For

This book is part of a series on Windows Azure service and client application development. However, it is not limited to only applications that run in Windows Azure. Windows Phone applications can interact with almost any service—they use data exposed by any on-premises or remote service. Even if you are building applications for Windows Phone that use other types of services (or no services at all), this book will help you to understand the Windows Phone environment, the development process, and the capabilities of the device.

This book is intended for any architect, developer, or information technology (IT) professional who designs, builds, or operates applications and services for Windows Phone. It is written for people who work with Microsoft® Windows-based operating systems. You should be familiar with the Microsoft .NET Framework, Microsoft Visual Studio® development system, and Microsoft Visual C#®. You will also find it useful to have some experience with Microsoft Expression Blend® design software and the Microsoft Silverlight® browser plug-in, although this is not a prerequisite.

Why This Book Is Pertinent Now

Mobile devices, and mobile phones in particular, are a part of the fundamental way of life for both consumers and business users. The rapidly increasing capabilities of these types of devices allow users to run applications that are only marginally less powerful, and in most cases equally (or even more) useful than the equivalent desktop applications. Typical examples in the business world are email, calendaring, document sharing, and other collaboration activities. In the consumer market, examples include access to social interaction sites, mapping, and games.

Windows Phone is a recent entry into this field, and it is very different from previous versions of Microsoft mobile operating systems. It has been built from the ground up to match the needs and aspirations of today's users, while standardizing the hardware to ensure that applications perform well on all Windows Phone devices. The result is a consistent run-time environment and a reliable platform that uses familiar programming techniques. In addition, the latest release of Windows Phone brings many new capabilities to the platform, enabling developers to create even better, more immersive user experiences.

Developers can use the tools they already know, such as Visual Studio, to write their applications. In addition, the Windows Phone 7.1 SDK provides a complete emulation environment and additional tools specially tailored for developing Windows Phone applications. Developers can use these tools to write, test, and debug their applications locally before they deploy them to a real device for final testing and acceptance. This book shows you how to use these tools in the context of a common scenario—extending an existing cloud-based application to Windows Phone.

How This Book Is Structured

You can choose to read the chapters in the order that suits your existing knowledge and experience, and select the sections that most interest you or are most applicable to your needs. However, the chapters follow a logical sequence that describes the stages of designing and building the application. Figure 1 illustrates this sequence.

- **Chapter 1, "The Tailspin Scenario,"** introduces you to the Tailspin company and the Surveys application. It describes the decisions that the developers at Tailspin made when designing their application, and it discusses how the Windows Phone client interacts with their existing Windows Azure-based services.
- **Chapter 2, "Building the Mobile Client,"** describes the steps that Tailspin took when building the mobile client application for Windows Phone that enables users to register for and download surveys, complete the surveys, and upload the results to the cloud-based service. It includes details of the overall structure of the application, the way that the Model-View-ViewModel (MVVM) pattern is implemented, and the way that the application displays data and manages commands and navigation between the pages. The following chapters describe the individual features of the application development in more detail.
- **Chapter 3, "Using Services on the Phone,"** discusses the way that the Windows Phone client application stores and manipulates data, manages activation and deactivation, uses live tiles, synchronizes data with the server application, and captures picture and sound data.

- **Chapter 4, "Connecting with Services,"** describes how the client application running on Windows Phone uses the services exposed by the Windows Azure platform. This includes user authentication, how the client application accesses services and downloads data, the data formats that the application uses, filtering data on the server, and the push notification capabilities.

The appendices cover unit testing Windows Phone applications, and information about the Prism Library that has been adapted for Windows Phone.

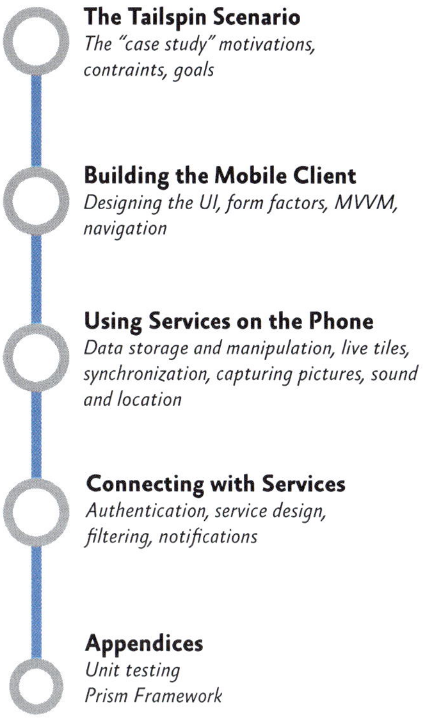

Figure 1
The book structure

The Example Application

This book has an accompanying example application—the Surveys client that Tailspin built to expose their cloud-based surveys application on Windows Phone. You can download the application and run it on your own computer to see how it works and to experiment and reuse the code.

The application is provided in two versions to make it easier for you to see just the Windows Phone client or the combined Windows Phone and Windows Azure application. If you want to try only the Windows Phone client, you can run the simplified version of the application that uses mock service implementations to provide the data required by the client application. You do not need to install the Windows Azure run-time environment and the Windows Azure emulator to use this version.

However, if you want to see the complete application in action, including features that require a back end (like push notifications), and work with the Windows Azure service, you can run the full version. For this, you must install the complete Windows Azure SDK and its run-time components. The example includes a dependency checker application that will assist you in identifying all the prerequisites, and get them installed and configured for this version; it will also help you locate and install any prerequisites that are missing on your system.

To read more and download the application, see *"A Case Study for Building Advanced Windows Phone Applications,"* on MSDN.

What You Need to Use the Code

These are the system requirements for running the scenarios:

- Microsoft Windows® Vista operating system (x86 and x64) with Service Pack 2 (all editions except Starter Edition) or Microsoft Windows 7 (x86 and x64) (all editions except Starter Edition)
- Microsoft Visual Studio 2010 Professional, Premium, or Ultimate edition
- Microsoft Visual Studio 2010 SP1

- Windows Phone 7.1 SDK
- Silverlight for Windows Phone Toolkit
- Microsoft Internet Information Services (IIS) 7.0

The Visual Studio solution uses features such as unit testing and folders, which are not currently available on Visual Studio Express.

To run the unit tests, you will also need:

- Silverlight unit test framework for Windows Phone
- Moq for .NET 4

If you want to run the full version of the sample, which uses a Windows Azure service to provide the data and authentication services to the device, you must also install the following:

- Windows Azure Tools for Microsoft Visual Studio 2010 (version 1.6)
- Windows Identity Foundation

Who's Who

This book uses a set of scenarios that demonstrate designing and building the Windows Phone client application and integrating it with cloud-based services. A panel of experts comments on the development efforts. The panel includes a Windows Phone specialist, a software architect, a software developer, and an IT professional. The scenarios can be considered from each of these points of view. The following table lists the experts for these scenarios.

Christine is a phone specialist. She understands the special requirements inherent in applications designed to be used on small mobile devices. Her expertise is in advising architects and developers on the way they should plan the feature set and capabilities to make the application usable and suitable for these types of devices and scenarios.

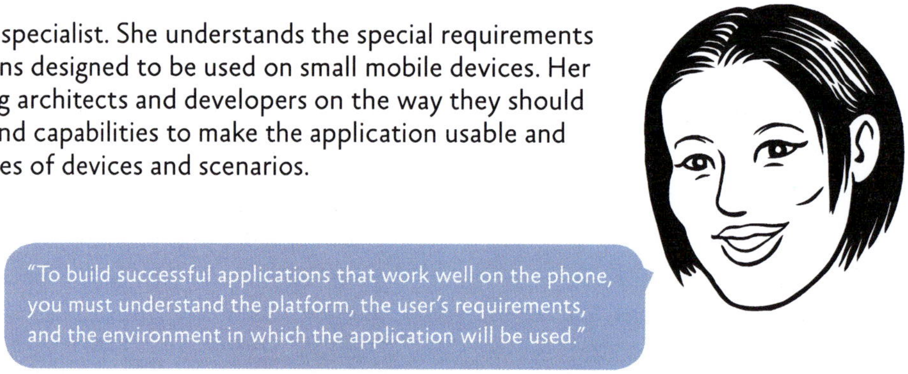

Jana is a software architect. She plans the overall structure of an application. Her perspective is both practical and strategic. In other words, she considers not only what technical approaches are needed today, but also what direction a company needs to consider for the future.

"It's not easy to balance the needs of the company, the users, the IT organization, the developers, and the technical platforms we rely on."

Markus is a senior software developer. He is analytical, detail-oriented, and methodical. He's focused on the task at hand, which is building a great application. He knows that he's the person who's ultimately responsible for the code.

"I don't care what platform you want to use for the application, I'll make it work."

Poe is an IT professional who's an expert in deploying and running applications in a corporate data center. Poe has a keen interest in practical solutions; after all, he's the one who gets paged at 3:00 AM when there's a problem.

"Integrating our server-based applications with mobile devices such as phones is a challenge, but it will broaden our reach and enable us to implement vital new capabilities for our applications and services."

If you have a particular area of interest, look for notes provided by the specialists whose interests align with yours.

Where to Go for More Information

There are a number of resources listed in text throughout the book. These resources will provide additional background, bring you up to speed on various technologies, and so forth. For your convenience, there is a bibliography online that contains all the links so that these resources are just a click away.

All links in this book are accessible from the book's online bibliography. You can find the bibliography on MSDN at: *http://msdn.microsoft.com/en-us/library/gg490786.aspx*.

1 The Tailspin Scenario

This chapter introduces a fictitious company named Tailspin. Tailspin's flagship product is an online service, named Surveys, that enables other companies or individuals to conduct their own online surveys. A year ago, Tailspin extended this service to mobile users, enabling subscribers to the Surveys application to publish surveys to people with Windows® Phone devices. These people used the Surveys mobile client application for Windows Phone OS 7.0 to capture survey data from the field, and it proved to be highly successful. Tailspin has now decided to develop a new version of the Surveys mobile client application that uses some of the new functionality available in Windows Phone OS 7.1, in order to improve the application experience. The chapters that follow show, step by step, how Tailspin designed and developed the Surveys mobile client application to run on Windows Phone devices that run Windows Phone OS 7.1.

The Tailspin Company

Tailspin is a two-year-old ISV company of approximately 30 employees that specializes in developing solutions using Microsoft® technologies. The developers at Tailspin are knowledgeable about various Microsoft products and technologies, including the .NET Framework, Windows Azure™ technology platform, Silverlight® browser plug-in, Microsoft Visual Studio® development system, and Windows Phone OS 7.0. The Surveys mobile client application was the first application that the developers at Tailspin created for the Windows Phone platform. It increased both the volume of survey responses, and their customer base. Tailspin believes that the time is right to develop a new mobile client application that will use some of the new functionality available in Windows Phone OS 7.1. They hope their innovative approach to collecting survey data that the mobile client application offers will continue to help it to grow its market share and increase its revenues.

Tailspin's Strategy

Tailspin is an innovative and agile organization; it is well placed to capitalize on new technologies, the business opportunities offered by the cloud, and the increasing sophistication of mobile phones. As an established company, Tailspin is willing to take risks and use new technologies when it implements applications.

Tailspin has gained a competitive advantage by being an early adopter of new technologies, especially in mobile devices and the cloud. It gained experience developing the Surveys mobile client application for Windows Phone OS 7.0, and will build on this experience when developing the Surveys mobile client application for Windows Phone OS 7.1.

Tailspin's Goals and Concerns

Tailspin wants to develop a new version of the Surveys mobile client application that uses some of the new functionality in Windows Phone OS 7.1.

The Surveys application has been a great success for Tailspin, and their market position was further improved by developing the Surveys mobile client application for Windows Phone OS 7.0. Tailspin first developed the mobile client application because subscribers wanted to be able to proactively find survey respondents. Instead of waiting for respondents to come to the survey website by following a link on a web page or in an email, subscribers wanted other ways of finding survey respondents. For example, subscribers wanted to be able to use surveyors who could go out and interview people.

When Tailspin developed the Surveys mobile client application for Windows Phone OS 7.0 they identified three key features they felt the Surveys application should have:

1. The application should support a wider range of question types and enable respondents to include additional data, such as pictures, audio, and location data, as a part of their survey responses.
2. It should allow people to provide survey responses when they are away from their computers. A convenient time to respond to a survey might not be a convenient time to be using a computer, for example during a commute or while waiting in a checkout line.
3. It should allow subscribers to capture a geographical location for the respondents answering a survey.

The developers at Tailspin already have Windows Phone OS 7.0 development skills.

The developers at Tailspin had a year's experience with Windows Phone as a platform, and were confident of their abilities to build the Surveys mobile client application for Windows Phone OS 7.1. However, the developers first had to understand the capabilities of Windows Phone OS 7.1 in order to determine how best to architect and design both the mobile client application and the elements of the application in the cloud.

There are many new features available in Windows Phone OS 7.1 including:

- Fast application switching
- Multitasking
- Background agents
- Background file transfers
- New sensors
- Network information
- Enhanced push notification support
- Live tiles
- Local database support
- Sockets support
- Encrypted credentials store
- Programmatic camera access

The Tailspin developers had to decide which of the new features would add value to the mobile client application and offer a better user experience.

Three key areas of concern for Tailspin in using the Windows Phone platform were reliability, security, and connectivity.

Windows Phone devices may be only intermittently connected to the Internet, so the mobile client application had to be capable of reliably storing the collected data until it could be sent to the cloud application. Tailspin also wanted to make sure that passwords held on the Windows Phone device were stored securely.

For some surveys, subscribers wanted to be able to determine the identity of the person submitting the survey data to the cloud application.

Tailspin also wanted to implement a service endpoint in Windows Azure that best supports the requirements of Windows Phone devices. The developers at Tailspin had to make decisions about the connectivity between the mobile client application and the back end, such as whether to use Representational State Transfer (REST)-style or SOAP-style web services, how "chatty" the interface should be, and how to handle retries when sending a message failed.

Finally, Tailspin wanted to be able to leverage the existing skills of its developers and minimize any retraining necessary to develop the Surveys mobile client application for Windows Phone OS 7.1.

A Windows Phone 7.5 device is a Windows Phone that is running Windows Phone OS 7.1.

The Surveys Application Architecture

Figure 1 shows a high-level view of the architecture of the Surveys application.

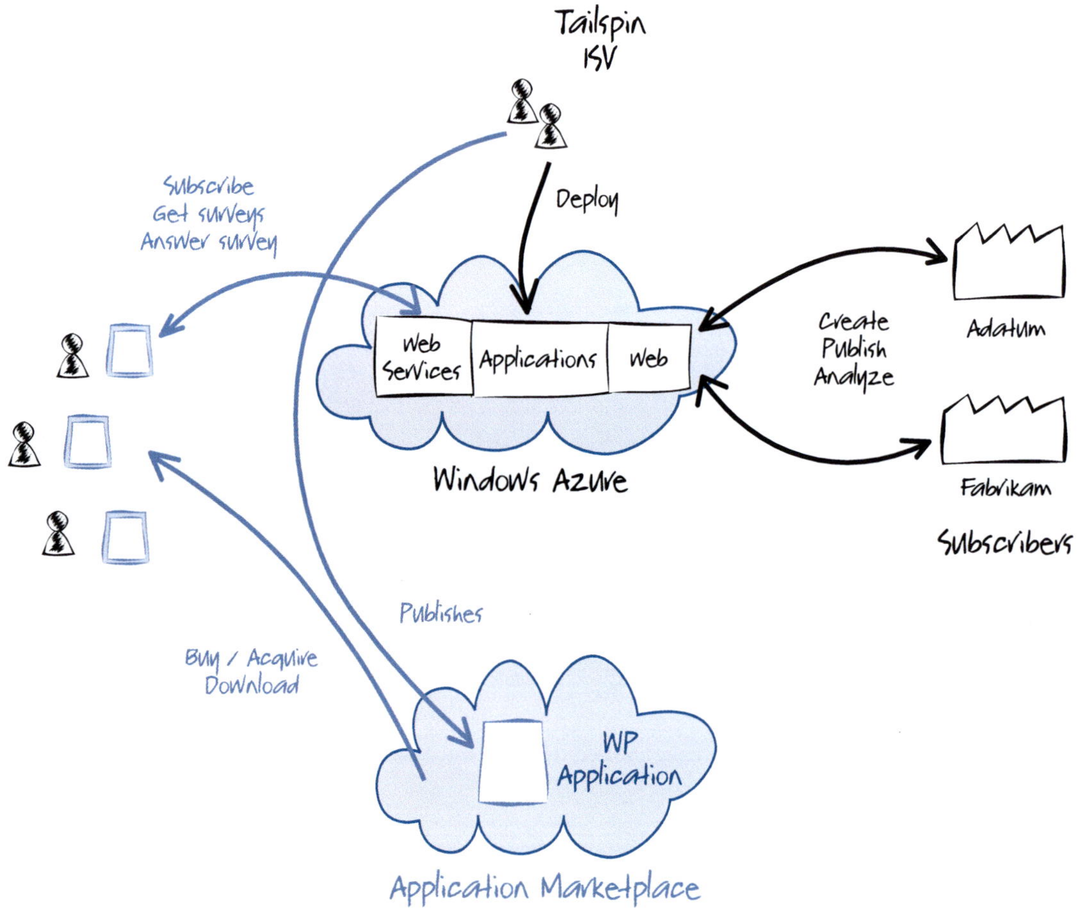

FIGURE 1
Architectural view of the Tailspin Surveys application

There are two, top-level components in the Surveys application. The first is the back end that Tailspin hosts in Windows Azure and that enables subscribers to create, publish, and analyze surveys. This back end is described in the book, *Developing Applications for the Cloud on the Microsoft Windows Azure Platform 2nd Edition*, available on the MSDN® website. The second component, which is the focus of this guidance, is a mobile client application that runs on Windows Phone devices and that enables surveyors to collect survey response data and send it to the back end. This guidance also describes the changes to the back-end cloud application that were necessary to support the mobile client application.

Tailspin is developing the mobile client application to support new features in the Surveys service. These new features include the following:

- The ability for surveyors to filter available surveys on different criteria.
- The ability to collect rich data from survey respondents, such as the respondent's location, voice recordings, and pictures, as part of the survey.
- The ability of the application to notify surveyors that new surveys are available.
- The ability of the application to download new surveys from the cloud service and upload survey answers to the cloud service, in the background.

The mobile client application will allow surveyors to filter surveys based on the tenant, but Tailspin could extend this in the future to include filters on factors such as survey length, target audience, and location.

The Actors

There are three actors in the scenario supported by the architecture: the ISV, the subscribers, and the surveyors.

Tailspin - The ISV

Tailspin has developed a multi-tenant, Software as a Service (SaaS) application named Surveys that it runs in the cloud. A range of subscribers—from individuals, through small companies, to large enterprises—uses the Surveys service to run custom surveys. Tailspin has also developed the mobile client application for Windows Phone devices running Windows Phone OS 7.1, described in this guidance, that it makes available to surveyors through the Windows Phone Marketplace.

Tailspin is taking advantage of features such as the camera and the GPS, which are part of the Windows Phone platform, to offer this functionality in the Surveys mobile client application.

Fabrikam and Adatum - The Subscribers

In the scenario, Fabrikam and Adatum are also fictitious companies who play the role of subscribers to the Surveys service. They design and launch surveys using the Surveys service, wait for responses, and then analyze the results that the Surveys application collects.

The Surveyors - Windows Phone Users

The surveyors, who typically work from home, subscribe to surveys based on a predefined criteria and are notified when new surveys are published. Using a Windows Phone device, they can either answer the survey questions themselves, or they can interview other people and use the device to capture the survey response data. For example, a surveyor could use the device to record traffic patterns at different times of the day or to go door-to-door collecting survey responses.

By using surveyors, Tailspin has targeted surveys more effectively and improved the response rate. Tailspin aims to further improve the application experience with the Surveys mobile client application for Windows Phone OS 7.1.

The Business Model

Tailspin's business model is to charge subscribers a monthly fee for access to the Surveys application, and Tailspin must then pay the actual costs of running the application. The Surveys mobile client application is free to surveyors, and surveyors who are collecting multiple responses to surveys can also be compensated. A Surveys subscriber such as Adatum, could either pay a surveyor for the number of submitted surveys or offer discount coupons. This works by identifying the surveyor who submitted the survey responses.

Tailspin will make the mobile client application available for free.

Tailspin is also planning to use the Microsoft Advertising SDK for Windows Phone to embed advertisements in the mobile client application as an additional way of generating revenue. The Advertising SDK is fully integrated into the Windows Phone 7.1 SDK and does not need to be installed separately.

> *The sample application that you can download to go with this guide doesn't implement any revenue-generating functionality; however, it is likely that a real-world version of the application would do this.*

The Application Components

Figure 2 illustrates the key functional components of the mobile client application and the relationships between them.

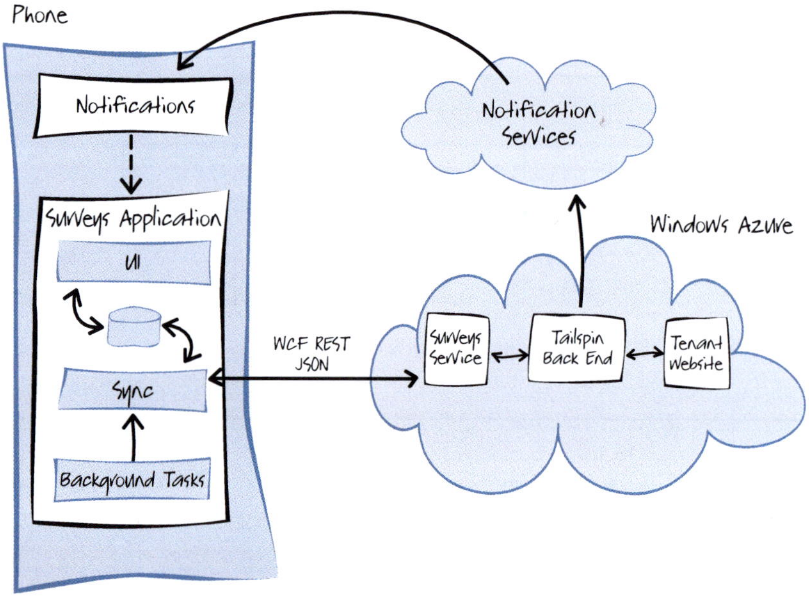

Figure 2
Tailspin Surveys, end to end

Developing Applications for the Cloud on the Microsoft Windows Azure Platform 2nd Edition describes the Tailspin back end and Subscriber website architecture, design, and implementation in detail. These components, which run on Windows Azure, enable subscribers to design new surveys and to analyze the responses that the application collects. The book also describes a public website that people can use to complete surveys using a web browser. The scenario described in this guidance focuses on an application running on Windows Phone OS 7.1 that provides an additional way for Tailspin to capture survey results.

The Windows Phone mobile client application is an alternative to using the web as a mechanism for collecting survey responses.

The Surveys application on Windows Phone comprises a number of components. A user interface (UI) enables the user to complete surveys and perform other tasks. A storage repository holds survey definitions and survey responses. A synchronization component is responsible for downloading survey definitions from the Tailspin back end and for uploading completed survey data.

To enable the Windows Phone application to communicate with the back end, the cloud components include an API that exposes the functionality that the mobile client application requires. Tailspin uses Windows Communication Foundation (WCF) REST to transport the data over the network. The Windows Phone application also authenticates with the back end so that the back end can determine which surveys it should make available to the mobile client and can track which responses come from which user. In the scenario described in this guidance, the mobile client application authenticates with the back end using basic authentication, but it is designed in such a way that it could be extended to accept more sophisticated mechanisms, such as a claims-based approach.

The application uses push notifications to inform the mobile client application that there are new surveys available to download. These push notifications will reach the Windows Phone device even when the Surveys mobile client application is not running.

Later chapters in this guidance describe these components in more detail.

All links in this book are accessible from the book's online bibliography. You can find the bibliography on MSDN at: *http://msdn.microsoft.com/en-us/library/gg490786.aspx*.

2 Building the Mobile Client

This chapter describes how the developers at Tailspin built the user interface (UI) components of the mobile client application. It begins by discussing some of the goals and requirements that Tailspin identified for the application before discussing, at a high level, the structure and key components of the application.

The chapter then discusses navigation and UI controls in more detail, and describes how and why Tailspin implemented the Model-View-ViewModel (MVVM) pattern. The chapter also gives an overview of the MVVM pattern itself.

The chapter includes discussions of the design Tailspin adopted for the application as well as detailed descriptions of the implementation.

Overview of the Mobile Client Application

This section provides an overview of the mobile client application to help you understand its overall structure before you examine the components that make up the application in more detail. Also, to help you understand some of the design decisions made by the developers at Tailspin, it describes some of the goals and requirements that Tailspin identified for the application.

Goals and Requirements

Windows® Phone OS 7.1 offers a wealth of features to developers and designers. The team at Tailspin wanted to ensure that their mobile client application makes the best possible use of the latest version of the platform and also plays by the rules. The application follows the recommended usability guidelines to ensure an optimal user experience and the "good citizen" best practices guidelines to ensure that the application makes efficient use of resources on the device in the context of the phone's functionality and other installed applications. They identified three sets of goals for the design and development of the application: usability goals, non-functional goals, and development process goals.

The Tailspin mobile client application follows usability and good citizen best practices.

For more information about Windows Phone UI design guidelines, see *User Experience Design Guidelines for Windows Phone* on the MSDN® developer program website.

Usability Goals

The usability goals are designed to ensure that the user's experience of the application meets her expectations for applications on a Windows Phone device. The following table lists some examples.

Goal Description	Example
Take advantage of the appearance and behavior of the Windows Phone platform.	The application accepts standard input gestures for users to enter data, uses the standard system colors, and includes icons designed to match the phone's theme. The application can update its appearance to blend with the phone's standard Light and Dark themes. For more information, see the **Themed-ResourceLocator** class in the Resources folder.
Use the standard controls Windows Phone users are familiar with.	The application uses standard controls, including the **ApplicationBar**, and **Pivot** controls, to make the user feel at home and to minimize the learning curve. In addition, the application also uses controls from the Microsoft Silverlight® for Windows Phone Toolkit.
Follow other Windows Phone UI guidelines, such as those on the use of the hardware Back button and the behavior of the application when the user answers a call or switches to another application.	The Back button navigates backward in a way that matches the user's expectations. The application restores the UI to its previous state after the user finishes answering an incoming call.
Integrate with the phone capabilities.	The application uses the location services on the phone to establish its geographical location, and it uses the camera and microphone to collect data for some survey questions.
Handle changes in screen orientation.	The application automatically updates the display orientation when the user changes the phone's orientation.
Handle standard screen resolutions.	The application displays correctly in the standard screen resolutions for Windows Phone devices.
The application should always have a responsive UI.	The application performs long-running tasks, such as synchronizing with the Surveys service, asynchronously or by the background agent. The application remains responsive when it has a large number of surveys saved locally and when it is displaying a survey with a large number of questions.
Take advantage of the background agent feature of the Windows Phone platform.	The data being used and produced by the application could synchronize with the back end in the cloud, even if the application is not running or being used.

Non-Functional Goals

The non-functional goals describe expected behaviors for the application, including some good citizen behaviors that relate to the limited resources on the device. The following table lists some examples.

Goal Description	Example
The application should continue to operate even when it is not connected to the back end in the cloud.	The application stores survey definitions and user responses in local storage, and it synchronizes with the cloud back-end store when connectivity is restored.
The application should not rely on specific network capabilities or assume a minimum available bandwidth.	The UI always interacts with local storage. The application uses an asynchronous call to synchronize with the Tailspin Surveys remote service and uses a store-and-forward pattern. Only the background agent, which uploads survey answers, requires a non-cellular connection and this is handled by the background agent itself.
The application should try to minimize the costs associated with using the network.	The application tries to minimize the amount of data transferred over the network by using JSON serialization instead of XML. The application does not compress the data because of the additional CPU overhead and battery consumption that this requires. In addition, the resource-intensive task, used by the background agent to upload survey answers, checks the current network interface type and only runs if a WiFi connection is available.
The application should proactively notify users of new information generated by the back end.	The back end uses the Microsoft Push Notification Service to notify users of new surveys available for their phones.
The application should use memory efficiently and, for performance, minimize memory allocations.	The sample application uses a dependency injection container to manage which objects are cached to improve performance and which objects are recreated whenever they are used.
As a "good citizen," the application should minimize its use of isolated storage—a shared resource on the phone.	The application removes completed surveys from isolated storage after the data successfully synchronizes with the Tailspin Surveys service. It also uses the JSON serializer when it saves data to isolated storage.

Users will prefer an application that fits well with the phone's UI design and theme. You will also have to comply with certain UI design guidelines if you want to distribute your application through Windows Marketplace.

Development Process Goals

Tailspin also identified a number of goals that relate to their own development processes. The following table lists some examples.

Goal Description	Example
Tailspin wants to have highly testable code.	A significant advantage of the MVVM pattern is that it makes the code more testable.
Tailspin wants to able to support other mobile platforms in the future.	Using standards-based approaches to interact with the back end makes it easier to develop other clients for other platforms.
Tailspin wants to have an efficient development process.	Developers and designers can work in parallel. Designers can prototype and build the UI using Microsoft Expression Blend® design software while the developers focus on the application's logic.
Tailspin wants to be able to adapt the application to work with any new capabilities of future versions of the Windows Phone platform.	The application uses an abstract persistence model to "wrap" local isolated storage on the device. Tailspin could easily change this in future to use the local database that resides in the application's isolated storage container.

The Components of the Mobile Client Application

Figure 1 shows the main components that comprise the Tailspin Surveys client application.

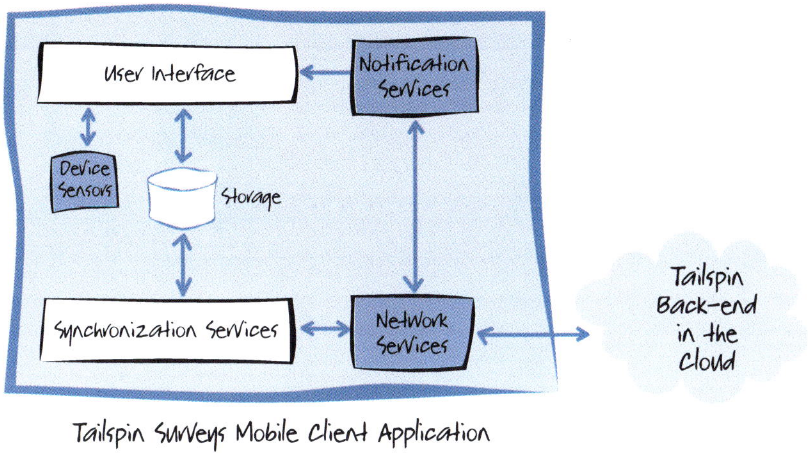

FIGURE 1
The Tailspin Surveys client application

The developers at Tailspin built three key components of the application: the UI, the storage sub-system, and the synchronization service. The application also uses some components of the Windows Phone platform; in particular, the GPS, camera, microphone, the notification services, and the network services that the application uses to communicate with the back-end web services.

The application uses a number of features offered by the Windows Phone platform.

This chapter focuses on the UI components and also describes how the application components are linked together through Tailspin's implementation of the MVVM pattern. Chapter 3, "Using Services on the Phone," will examine the storage and synchronization components, and Chapter 4, "Connecting with Services," will look at the notification process and the integration with the back end in more detail.

The Structure of the Tailspin Surveys Client Application

Figure 2 shows the structure of the Tailspin Surveys mobile client application in more detail. For clarity, the diagram does not show all the links between all the components. In particular, multiple links exist between the model components and the view model and the application services, but showing all of these would unnecessarily clutter the diagram.

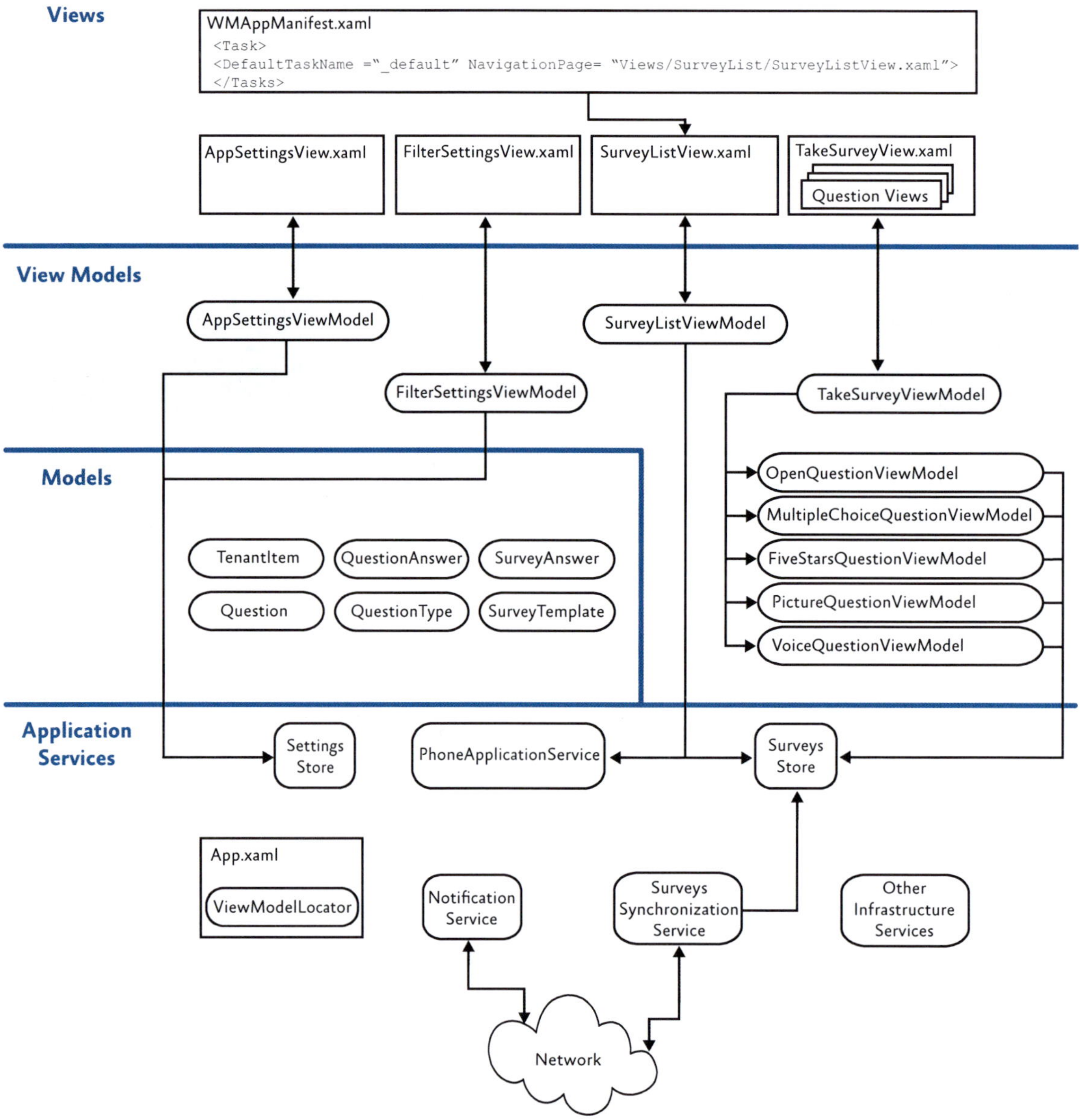

FIGURE 2
Tailspin Surveys mobile client application structure

To understand how Tailspin built the UI components (such as the **SurveyListView** page and the **AppSettingsView** page), how the navigation between the pages work, and how the application determines which page to display to the user when the user launches the application, you should read the section, "The Design of the User Interface," later in this chapter.

To understand how and why Tailspin uses the MVVM pattern, you should read the section, "Using the Model-View-ViewModel Pattern," later in this chapter. This section explains the roles of the view, view model, and model components and how they are linked together, including the role of the **ViewModelLocator** class. This section also describes some data-binding scenarios in the application, including the way the application uses the **Pivot** control on the **SurveyListView** page and on the **TakeSurveyView** page.

To understand how the application manages its state when it's dormant or tombstoned, you should read the section, "Handling Activation and Deactivation," in Chapter 3, "Using Services on the Phone."

*An application is made dormant by the Windows Phone device when, for example, the user navigates to another application or answers a call while using the application. In this state, the application remains intact in memory but no processing takes place. If the application is reactivated from this state, it does not need to recreate any state because it has been preserved. Dormant applications may be tombstoned by the operating system in order to free up memory. A tombstoned application has been terminated, but information about its navigation state and state dictionaries are preserved for when the application is relaunched. A device will maintain tombstoning information for up to five applications at once. For more information see, **"Execution Model Overview for Windows Phone,"** on MSDN.*

Dependency injection enables decoupling of concrete types from the code that depends on these types. It uses a container that holds a list of registrations and mappings between interfaces and abstract types and the concrete types that implement or extend these types.

To understand how the application manages persistent data on the phone, such as application settings and survey responses, you should read the section, "Using Isolated Storage on the Phone," in Chapter 3, "Using Services on the Phone."

To understand how the Tailspin Surveys cloud application can notify the mobile client of new surveys by using the push notification service, you should read Chapter 4, "Connecting with Services."

To understand how the application transfers survey data between the mobile client application and the cloud application, you should read Chapter 4, "Connecting with Services."

Dependency Injection

The developers at Tailspin use a dependency injection container to manage the instantiation of many of the classes, including the view model classes.

Tailspin uses the *Funq* dependency injection container instead of the Unity Application Block (*Unity*) because Unity is not available for the Windows Phone platform. The Funq dependency injection container is also lightweight and very fast.

The **ContainerLocator** class shows how the application creates the registrations and mappings in the Funq dependency injection container. In the Tailspin mobile client application, the **ViewModelLocator** instantiates the **ContainerLocator** object and is the only class in the application that holds a reference to a **ContainerLocator** object.

By default, the Funq dependency injection container registers instances as shared components. This means that the container will cache the instance on behalf of the application, with the lifetime of the instance then being tied to the lifetime of the container.

You should consider carefully which objects you should cache and which you should instantiate on demand. Caching objects improves the application's performance at the expense of memory utilization.

The TailSpin Solution

The TailSpin.PhoneOnly solution organizes the source code and other resources into projects. The following table outlines the main projects that make up the Surveys mobile client application.

Project	Description
TailSpin.Phone.Adapters	This project contains interfaces, adapters, and facades for Windows Phone API functionality.
TailSpin.PhoneAgent	This project contains a background agent implementation that launches a periodic and a resource-intensive task.
TailSpin.PhoneClient	This project contains the views and view models for the Surveys mobile client application, along with supporting classes and resources.
TailSpin.PhoneClient.Adapters	This project contains interfaces, adapters, and facades for Windows Phone API functionality that is not supported by background agents. Creation of this project was necessary in order to pass the capability validation performed as part of the Windows Phone Marketplace application submission process. This is because the set of APIs not supported by background agents must reside in a project not referenced by the TailSpin.PhoneAgent project. For more information, see "*Unsupported APIs for Background Agents for Windows Phone*" on MSDN.
TailSpin.PhoneServices	This project contains web service client implementations that interact with the Tailspin Surveys service in the cloud.

The Contents of the TailSpin.PhoneClient Project

The TailSpin.PhoneClient project organizes the source code and other resources into folders. The following table outlines what is contained in each folder and provides references to where this guide describes the content in more detail.

Project	Description
Root	In the root folder of the project, you'll find the App.xaml file that every Microsoft Silverlight® project must include. This defines some of the startup behavior of the application. The root folder also contains some image files that all Windows Phone applications must include.
Properties	In this folder, you'll find two manifest files and the AssemblyInfo.cs file. The WMAppManifest.xml file describes the capabilities of the application and specifies the initial screen to display when the user launches the application. In addition, it also contains the details of the background agent used by the application.
Resources	This folder holds various image files that the application uses and a utility class that performs conversions to types that are used in UI.
Views	This folder contains the views that define the screens in the application. The section "Using the Model-View-ViewModel Pattern" in this chapter describes the role of views in this pattern and highlights the fact that there should be little or no code in the code-behind files.
ViewModels	This folder contains the view models. The section "Using the Model-View-ViewModel Pattern" in this chapter describes the role of view models in this pattern. You will find more view models than views because individual user controls may also have their own view models.
Themes	The XAML file in this folder contains style definitions.
Services	This folder contains the **ContainerLocator** class, which creates the registrations and mappings in the Funq dependency injection container. The folder also contains the **ScheduledActionClient** class that uses the **ScheduledActionService** class from the Windows Phone API.
Services/RegistrationService	This folder contains client implementations that interact with the Tailspin Surveys service in the cloud. Chapter 4, "Connecting with Services," describes the web service clients in this folder.
Infrastructure	This folder contains utility code required by the application. The section "Using XNA Interop to Record Audio" in Chapter 3, "Using Services on the Phone," describes how some of the classes in this folder are used.

The Contents of the TailSpin.PhoneClient.Adapters Project

This project contains interfaces that mirror functionality in several classes in the Windows Phone 7.1 SDK, including **ICameraCaptureTask**, **IShellTile**, and **INavigationService**.

This project also includes adapter implementations of these interfaces that simply pass parameters to and return values from the underlying instances of the SDK classes. These interfaces and adapters are used in the mobile client application to increase testability. For more information see Appendix A, "Unit Testing Windows Phone Applications."

This project is similar to the TailSpin.Phone.Adapters project. Both projects have interfaces and adapters that "wrap" functionality in the Windows Phone 7.1 SDK. These two projects are separated because the set of APIs not supported by background agents must reside in a project not referenced by the TailSpin.PhoneAgent project. The TailSpin.PhoneAgent project utilizes the adapters in the TailSpin.Phone.Adapters project but not the TailSpin.PhoneClient.Adapters project.

The Contents of the TailSpin.PhoneServices Project

The TailSpin.PhoneServices project organizes the source code into folders. The following table outlines what is contained in each folder and provides references to where this guide describes the content in more detail.

Project folder	Description
Models	This folder contains the models. The section "Using the Model-View-ViewModel Pattern" in this chapter describes the role of the models.
Services/Clients	This folder contains the HttpClient class that makes asynchronous web requests. Chapter 4, "Connecting with Services," describes this class.
Services/RegistrationService Services/SurveysService	These folders contain client implementations that interact with the Tailspin Surveys service in the cloud. Chapter 4, "Connecting with Services," describes the web service clients in this folder.
Services/Stores	This folder contains classes for persisting application settings and survey data to and from isolated storage. Chapter 3, "Using Services on the Phone," describes these stores.

The Design of the User Interface

User Experience Design Guidelines for Windows Phone describes best practices for designing the UI of a Windows Phone application.

The Tailspin Surveys mobile client application follows the UI design guidance published in the *User Experience Design Guidelines for Windows Phone,* which you can view on MSDN.

This section describes how users navigate between the different pages of the mobile client application and outlines the controls that Tailspin uses in the application's UI.

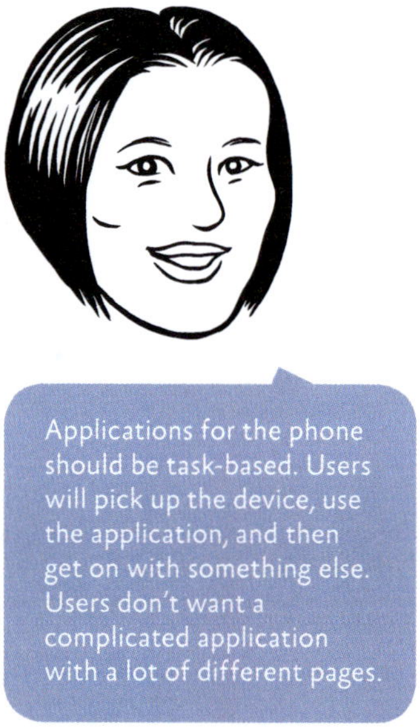

Page Navigation

The Tailspin mobile client application uses only a small number of pages, with a limited number of navigation routes between those pages. However, it also supports pinning Tiles to Start. A Tile is a link to an application displayed in Start. An Application Tile can be pinned to Start, which when tapped by the user will launch the application. Users can also pin a survey as a secondary Tile to Start. When a secondary Tile is tapped by the user, the application is launched and the survey page is navigated to. The intention of this is to offer the user quick and easy access to part of the application. For more information about tiles, you should read the section, "Using Live Tiles on the Phone," in Chapter 3, "Using Services on the Phone."

Figure 3 shows how the user navigates within the application on the phone.

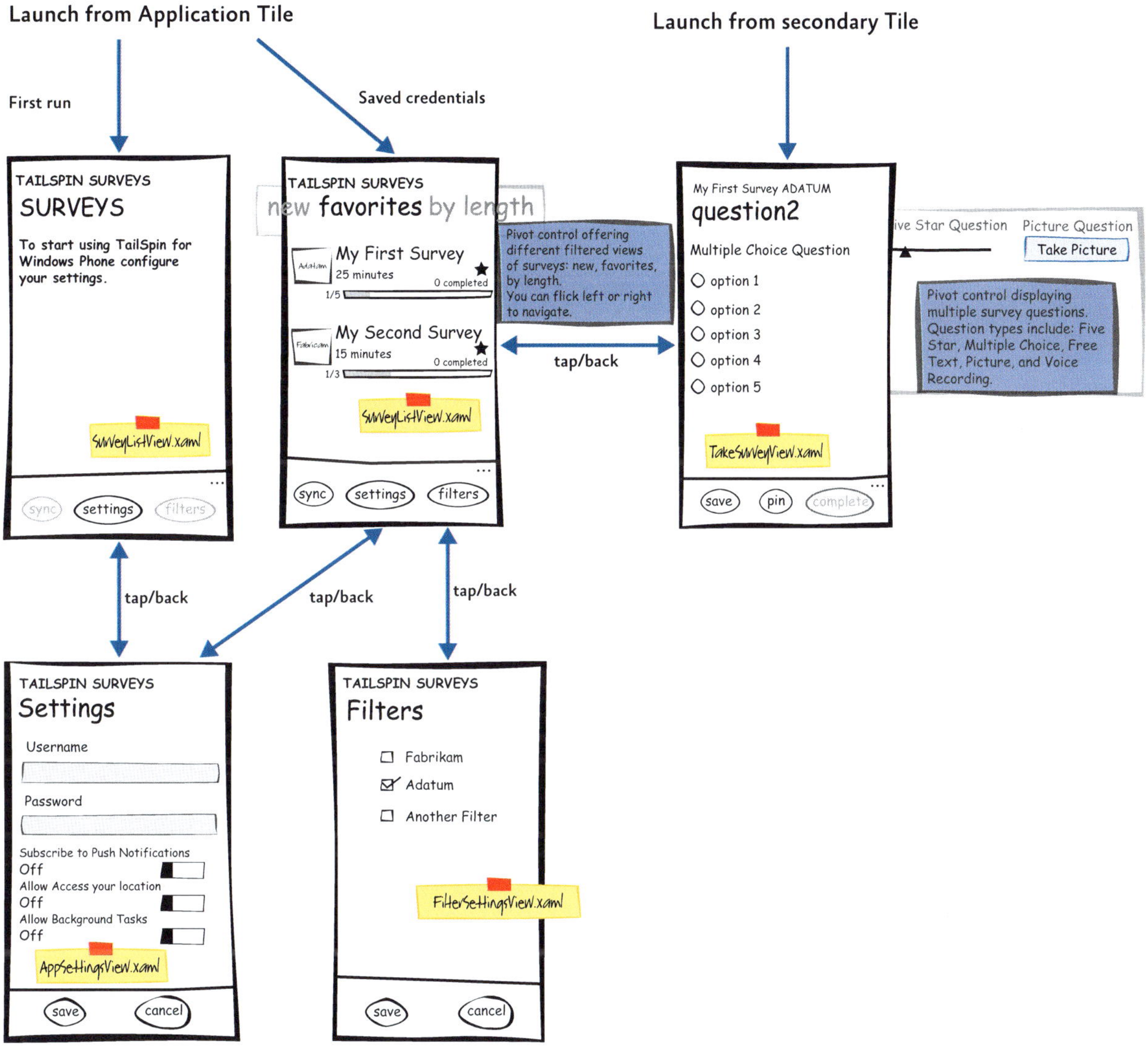

FIGURE 3
Navigation in the Surveys client application

When the application is launched from anything other than a secondary Tile, the initial screen to display is determined from the **DefaultTask** element in the WMAppManifest.xml file. The following code example shows how the application is configured to first display the SurveyListView page to the user.

```xml
XML
<Tasks>
  <DefaultTask Name ="_default"
    NavigationPage="Views/SurveyList/SurveyListView.xaml"/>
</Tasks>
```

When the application is launched from a secondary Tile, the application launches the page specified by the navigation destination of the Tile.

Before users can use the application for the first time, they must enter the credentials that will be used when the application synchronizes with the Tailspin Surveys service. The developers at Tailspin considered automatically navigating users to the AppSettingsView page if they haven't already supplied their credentials, but this introduced an issue with the way navigation behaves in the application. If the application automatically navigates the user to the AppSettingsView page from the SurveyListView page, and if the user then decides he or she doesn't want to enter credentials (maybe the application was started by mistake), the user will press the Back button and expect to leave the application. A simple approach to navigating will have left the SurveyListView page on the navigation stack, so that's where the user will end up. For some possible solutions to this problem, see the post, *"Redirecting an initial navigation,"* on Peter Torr's blog.

The application does not automatically navigate users to the **AppSettingsView** page; instead, it displays a message that explains to users that they must provide their credentials. The following code example from the SurveyListView.xaml file shows how the visibility of the message is controlled based on the value of the **SettingAreNotConfigured** property.

```xml
XAML
<StackPanel x:Name="SettingNotConfiguredPanel" Grid.Row="0"
  Margin="12,17,0,28"
  Visibility="{Binding SettingAreNotConfigured,
  Converter={StaticResource VisibilityConverter}}">
  <TextBlock x:Name="ApplicationTitle" Text="TAILSPIN SURVEYS"
    Style="{StaticResource PhoneTextNormalStyle}"/>
  <TextBlock x:Name="PageTitle" Text="Surveys" Margin="9,-7,0,0"
    Style="{StaticResource PhoneTextTitle1Style}"/>
  <ContentControl Template="{StaticResource
    SettingsNotConfiguredTextBlock}" />
</StackPanel>
```

The following code example from the Styles.xaml file shows the template that defines the message.

```xml
XAML
<ControlTemplate x:Key="SettingsNotConfiguredTextBlock">
  <TextBlock
    VerticalAlignment="Top"
```

```
    Margin="12"
    Style="{StaticResource PhoneTextLargeStyle}"
    Foreground="{StaticResource PhoneSubtleBrush}"
    Text="To start using TailSpin for Windows Phone, configure
            your Settings."
    TextWrapping="Wrap"/>
</ControlTemplate>
```

When you navigate using the **NavigationService** class, the behavior of the Back button is automatically determined, so using the Back button on the SurveyListView page causes the application to exit, and using the Back button on the AppSettingsView page returns the user to the SurveyListView page.

*The **NavigationService** class automatically manages the behavior of the Back button.*

The following code example from the **AppSettingsViewModel** class shows how the application implements the navigation away from the AppSettingsView page in code for the Cancel button. Notice how the **Cancel** method uses the **NavigationService** class. This class is described in detail later in this chapter in the section, "Handling Navigation Requests."

C#

```
public void Cancel()
{
  if (this.NavigationService.CanGoBack)
    this.NavigationService.GoBack();
}
```

When taking a survey, there are three scenarios that could be invoked to navigate away from the TakeSurveyView page. First, the user could press the back button on the phone. Second, the user could click the complete survey button on the application bar. Third, the user could click the save survey button on the application bar. The following code example from the **TakeSurveyViewModel** class shows how the application implements these scenarios to navigate away from the TakeSurveyView page. Notice how the **CleanUpAndGoBack** method tests the **CanGoBack** property before calling the **GoBack** method from the **NavigationService** class. This is because users could have arrived at the page from either the SurveyListView page or from a pinned tile. If users have arrived at the page from the SurveyListView page, the **CanGoBack** property will be true and thus the **GoBack** method can be called. If users have arrived at the page from a pinned tile, the **CanGoBack** property will be false, preventing the **GoBack** method from being called.

C#
```
private void CleanUpAndGoBack(bool completed)
{
  if (NavigationService.CanGoBack)
  {
    ...
    this.NavigationService.GoBack();
  }
  else if (completed)
  {
    ...
  }
  else
  {
    ...
  }
}
```

Navigating from the survey list screen to an individual survey is a little more complicated because the application must display the survey that the user currently has selected in the list. Furthermore, the application must respond to the user tapping on an item in a **ListBox** control, on a **PivotItem** in a **PivotControl**.

When the user taps a survey name in the list, the navigation to the TakeSurveyView page is accomplished using an interaction trigger in the **SurveyDataTemplate** data template. The following code example from the Styles.xaml page shows this data template definition along with the style for the **ListBox ItemTemplate**.

XAML
```
<DataTemplate x:Key="SurveyDataTemplate">
  ...
  <i:Interaction.Triggers>
    <i:EventTrigger EventName="MouseLeftButtonUp">
      <i:InvokeCommandAction Command="{Binding TakeSurveyCommand}"/>
    </i:EventTrigger>
  </i:Interaction.Triggers>
  ...
</DataTemplate>

<!--Style for Survey List-->
<Style x:Key="SurveyTemplateItemStyle" TargetType="ItemsControl">
  <Setter Property="ListBox.ItemTemplate"
    Value="{StaticResource SurveyDataTemplate}"/>
</Style>
```

The action that enables navigation from a tap on an item in the **ListBox** control is provided by the **EventTrigger** and **InvokeCommandAction** class. This action binds a command to a user tap on an element in the view—in this example, a tap on an item in a list to the **TakeSurveyCommand** command in the **SurveyTemplateViewModel** view model class. For an explanation of how the TakeSurveyView page displays the correct survey, based on the survey selected in the list, see the section, *"Connecting the View and the View Model,"* later in this chapter.

User Interface Description

Figure 3, earlier in this chapter, shows a mockup of the UI and the navigation routes supported by the application between the pages. There are a few items in the UI that require some additional explanation.

The SurveyListView page displays the following information about each survey:

- The survey creator's logo.
- The survey's title.
- A number to indicate the number of completed surveys pending upload.
- The survey creator's name.
- A star to indicate whether the survey is one of the user's favorites.
- A number to indicate how many questions have been answered so far.
- A progress bar to indicate how many questions have been answered so far.
- A number to indicate the estimated amount of time it should take to complete the survey.

On the TakeSurveyView page, there are three buttons on the application bar. The **Save** button saves the current answers to the survey and allows the user to return to the survey later to amend existing answers and to add additional answers. The **Pin** button adds a secondary Tile to the Start screen with the application icon, and with the survey name as the title. Tapping the title sends the user directly to the survey in the application. The **Complete** button saves the current answers and marks the survey as complete and ready for synchronization. You cannot return to or change a completed survey.

User Interface Elements

The application uses standard controls throughout so that the appearance and behavior of the application matches the Windows Phone standard appearance and behavior. The section, "Displaying Data," later in this chapter describes how the application implements data binding with the Pivot control.

The Pivot Control

The application uses a **Pivot** control on the SurveyListView page to enable the user to view different filtered lists of surveys, such as new surveys, or favorite surveys, or the list of surveys sorted by length. The control allows the user to navigate between the different lists by panning left or right, or by using flick gestures in the application in a way that is consistent with the user's expectations in the Windows Phone UI. The developers at Tailspin chose to use the **Pivot** control here because it enabled them to display a set of items that all have the same data type: in Tailspin Surveys, each **PivotItem** control displays a list of surveys.

The application also uses a **Pivot** control on the TakeSurveyView page to display the survey questions and collect the survey responses. The developers at Tailspin chose to use the **Pivot** control here because its large scrollable area offers a great way to display a complete survey and long text items, and offers an intuitive way to navigate by scrolling left or right through the questions.

Styling and Control Templates

The file Styles.xaml in the Themes folder contains some styling information for several of the controls used on the pages in the application. The **ListBox** controls that the application uses to display lists of surveys on the SurveyListView page uses the **SurveyTemplateItemStyle** style.

The SurveyListView page uses the **NoItemsTextBlock** control template to display a message when there are no surveys to display in the **ListBox** control.

The SurveyListView page uses the **SettingNotConfiguredPanel** control template to display a message when the user hasn't configured his settings in the application.

The **ThemedResourceLocator** class in the Resources folder shows how the application handles UI changes if the user chooses either the Dark or Light Windows Phone themes. Although most controls automatically adjust to different UI themes, there are a few places in the Tailspin mobile client that need some additional logic, such as where the application uses custom icons.

Context Menus

On the SurveyListView page, if the user taps a survey name, the application navigates to the Take-SurveyView page and displays the survey questions. If the user taps and holds on a survey name on the SurveyListView page, the application displays a context menu.

> *You can find the behavior that causes the context menu to display in the* ***ContextMenu*** *control itself.*

Tailspin uses the **ContextMenu** control from the Silverlight for Windows Phone Toolkit, which is available on the *Silverlight Toolkit* page on CodePlex.

The following code example from the data template in the Styles.xaml file shows how Tailspin declares the **ContextMenu** control that displays the context menu when the user taps and holds on a survey name.

XAML

```
<toolkit:ContextMenuService.ContextMenu>
  <toolkit:ContextMenu>
    <toolkit:MenuItem Header="mark as favorite"
      Command="{Binding MarkFavoriteCommand}"
      Visibility="{Binding IsFavorite,
      Converter={StaticResource NegativeVisibilityConverter}}"/>
    <toolkit:MenuItem Header="remove mark as favorite"
      Command="{Binding RemoveFavoriteCommand}"
      Visibility="{Binding IsFavorite,
      Converter={StaticResource VisibilityConverter}}"/>
    <toolkit:MenuItem Header="pin to start"
      Command="{Binding PinCommand}"
      IsEnabled="{Binding IsPinnable}"/>
  </toolkit:ContextMenu>
</toolkit:ContextMenuService.ContextMenu>
```

Using the Model-View-ViewModel Pattern

Now that you've seen how Tailspin designed the UI of the application, including the choice of controls, and the navigation through the application, it's time to look behind the scenes to discover how Tailspin structured the mobile client for the Windows Phone platform.

The developers at Tailspin decided to adopt a Model-View-ViewModel (MVVM) approach to structuring the Windows Phone application. This section provides an overview of MVVM, explains why it is appropriate for Windows Phone applications, and highlights some of the decisions made by the developers at Tailspin about the implementation.

The application is designed and built using the MVVM pattern.

The Premise

The MVVM approach naturally lends itself to XAML application platforms such as Silverlight. This is because the MVVM pattern leverages some of the specific capabilities of Silverlight, such as data binding, commands, and behaviors. MVVM is similar to many other patterns that separate the responsibility for the appearance and layout of the UI from the responsibility for the presentation logic; for example, if you're familiar with the Model-View-Controller (MVC) pattern, you'll find that MVVM has many similar concepts.

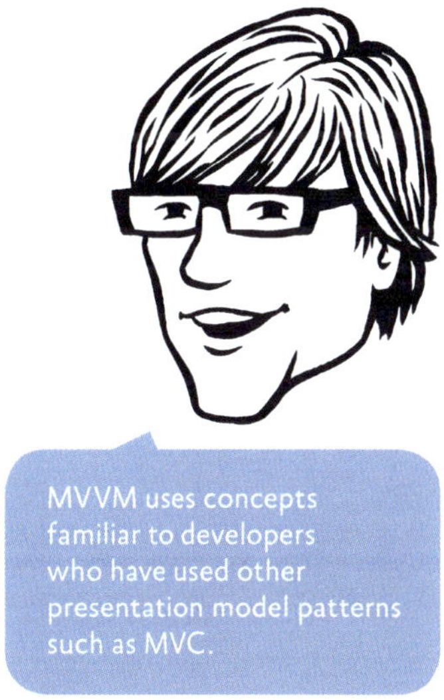

Overview of MVVM

There are three core components in the MVVM pattern: the model, the view, and the view model. Figure 4 illustrates the relationships between these three components.

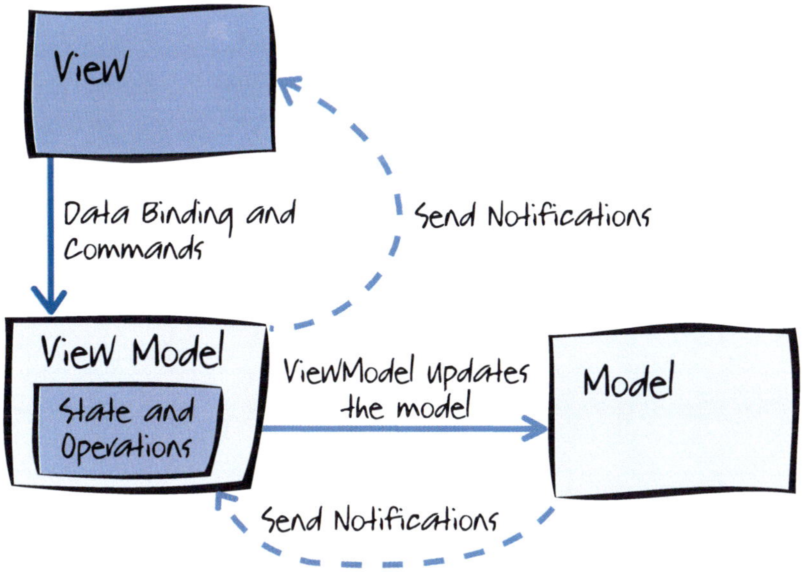

Figure 4
The Model-View-ViewModel pattern

In some scenarios, the view model may call a web service directly instead of using a model class that itself makes a call to the web service. For example, if a web service retrieves a collection of **Person** objects that you can deserialize and use directly in the view model, there's no need to define another **Person** class in the model.

The view model isolates the view from the model classes and allows the model to evolve independently of the view.

The view is responsible for defining the structure, layout, and appearance of what the user sees on the screen. Ideally, the view is defined purely with XAML, with a limited code-behind that does not contain business logic.

The model in MVVM is an implementation of the application's domain model that can include a data model along with business and validation logic. Often, the model will include a data access layer. In the case of a Windows Phone application, the data access layer could support retrieving and updating data by using a web service or local storage.

The view model acts as an intermediary between the view and the model, and is responsible for handling the view logic. The view model provides data in a form that the view can easily use. The view model retrieves data from the model and then makes that data available to the view, and may reformat the data in some way that makes it simpler for the view to handle. The view model also provides implementations of commands that a user of the application initiates in the view. For example, when a user clicks a button in the UI, that action can trigger a command in the view model. The view model may also be responsible for defining logical state changes that affect some aspect of the display in the view, such as an indication that some operation is pending.

In addition to understanding the responsibilities of the three components, it's also important to understand how the components interact with each other. At the highest level, the view "knows about" the view model, and the view model "knows about" the model, but the model is unaware of the view model, and the view model is unaware of the view.

MVVM leverages the data-binding capabilities in Silverlight to manage the link between the view and view model, along with behaviors and event triggers. These capabilities limit the need to place business logic in the view's code-behind.

Typically, the view model interacts with the model by invoking methods in the model classes. The model may also need to be able to report errors back to the view model by using a standard set of events that the view model subscribes to (remember that the model is unaware of the view model). In some scenarios, the model may need to be able to report changes in the underlying data back to the view model; again, the model should do this using events.

This chapter focuses on the view and view model components of the Tailspin mobile client application. Chapter 3, "Using Services on the Phone," includes a description of the model components in the application.

Benefits of MVVM

MVVM enables a great developer-designer workflow that promotes these benefits:

- During the development process, developers and designers can work more independently and concurrently on their components. The designers can concentrate on the view, and if they are using Expression Blend, they can easily generate sample data to work with, while the developers can work on the view model and model components.
- The developers can create unit tests for the view model and the model without using the view. The unit tests for the view model can exercise exactly the same functionality as used by the view.
- It is easy to redesign the UI of the application without touching the code because the view is implemented entirely in XAML. A new version of the view should work with the existing view model.
- If there is an existing implementation of the model that encapsulates existing business logic, it may be difficult or risky to change. In this scenario, the view model acts as an adapter for the model classes and enables you to avoid making any major changes to the model code.
- Although the benefits of MVVM are clear for a complex application with a relatively long shelf life, the additional work needed to implement the MVVM pattern may not be worthwhile for simple applications or applications with shorter expected lifetimes.

You should evaluate carefully whether MVVM is appropriate for your application, considering the initial overhead of using this approach.

You've seen a high-level description of the MVVM pattern, and the reasons that Tailspin decided to adopt it for their Windows Phone client. The next sections describe the following implementation choices made by the developers at Tailspin when they implemented MVVM for the Surveys application:

- How Tailspin connects the view and the view model components.
- Examples of how Tailspin tests components of the application
- How Tailspin implements commands, asynchronous operations, and notifications to the user.
- Details of data binding and navigation.

For more information about MVVM, see Chapter 5, *"Implementing the MVVM Pattern,"* and Chapter 6, *"Advanced MVVM Scenarios,"* of the *Prism* documentation on MSDN.

Connecting the View and the View Model

Now is a good time to walk through the code that implements the MVVM pattern in the Windows Phone application in more detail. As you read through this section, you may want to download the *Windows Phone Tailspin Surveys* application.

There are several ways to connect the view model to the view, including direct relations and data template relations. The developers at Tailspin chose to use a view model locator because this approach means that the application has a single class that is responsible for connecting views to view models. This still leaves developers free to choose to manually perform the connection within the view model locator, or by using a dependency injection container. Figure 5 illustrates the relationships between the view, view model locator, and container locator.

Tailspin adopted the view model locator approach to connecting the view to a view model. This approach works well for applications with a limited number of screens (which is often the case with Windows Phone applications), but is not always appropriate in larger applications.

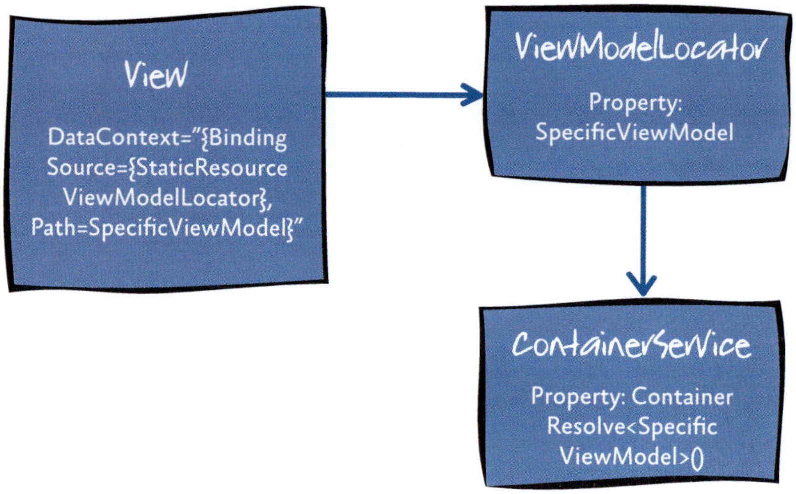

FIGURE 5
Connecting the view to the view model

Inside the Implementation

The mobile client application uses a view model locator to connect view models to views. The following code example from the App.xaml file shows how the view model locator class is identified and made available to the application. The application declares the **TailSpin.PhoneClient.ViewModels ViewModelLocator** class in the <Application.Resources> section of the App.xaml file.

```
XAML
<Application
    x:Class="TailSpin.PhoneClient.App"
    ...
    xmlns:viewmodels=
      "clr-namespace:TailSpin.PhoneClient.ViewModels"
    ... >

  <!--Application Resources-->
  <Application.Resources>
    ...
    <viewmodels:ViewModelLocator x:Key="ViewModelLocator"/>
    ...
  </Application.Resources>
  ...
</Application>
```

The following code example from the SurveyListView.xaml file shows how a view can then reference the **ViewModelLocator** class as a static resource in its data context bindings.

```
XAML
<phone:PhoneApplicationPage
    x:Class="
      TailSpin.PhoneClient.Views.SurveyList.SurveyListView"
    ...
    DataContext=
      "{Binding Source={StaticResource ViewModelLocator},
        Path=SurveyListViewModel}"
    ...
    >
    ...
</phone:PhoneApplicationPage>
```

The **Path** attribute identifies the property of the **ViewModelLocator** instance that returns the view model associated with the current page.

The following code example shows the parts of the **ViewModelLocator** class that return the **SurveyListViewModel** instance.

The ViewModelLocator class connects view models to views.

```
C#
public class ViewModelLocator : IDisposable
{
  private readonly ContainerLocator containerLocator;
  ...

  public SurveyListViewModel SurveyListViewModel
```

```
  {
    get
    {
      return this.containerLocator.Container
        .Resolve<SurveyListViewModel>();
    }
  }
}
```

Notice how the instance of the **ContainerLocator** class exposes the Funq **Container** property to resolve the view model.

> *The **ContainerLocator** class is also responsible for instantiating the store and passing it to the view model, which in turn passes it on to the model.*

Testing the Application

One of the benefits of combining the MVVM pattern with the dependency injection pattern is that it promotes the testability of the application, making it easy to create tests that exercise the view model.

Inside the Implementation

Tailspin uses a Silverlight unit-testing framework to run unit tests on the phone emulator and on real devices.

Tailspin uses the Silverlight unit test framework for Windows Phone and Silverlight 4.

The Windows Phone project named Tailspin.PhoneClient.Tests contains the unit tests for the Surveys mobile client application. To run the tests, first build and then deploy this project either to the Windows Phone emulator or to a real device. On the Windows Phone device, you can now launch an application named Tailspin.Tests, and then select the unit tests you want to run.

The following code example shows a unit test method from the **SurveyListViewModelFixture** class that tests that the view model returns a list of all the surveys that are marked as favorites.

C#
```
[TestMethod]
public void FavoritesSectionShowsFavoritedItems()
{
  var store = new SurveyStoreMock();
  var surveyStoreLocator = new SurveyStoreLocatorMock(store);
  store.Initialize();
  var allSurveys = store.GetSurveyTemplates();

  var vm = new SurveyListViewModel(surveyStoreLocator,
    new SurveysSynchronizationServiceMock(),
    new MockNavigationService(),
```

```
    new MockPhoneApplicationServiceFacade(),
    new MockShellTile(),
    new MockSettingsStore(),
    new MockLocationService());
  vm.Refresh();

  var favoriteSurveys =
    vm.FavoriteItems.Cast<SurveyTemplateViewModel>().ToList();
  CollectionAssert.AreEquivalent(
    allSurveys.Where(p => p.IsFavorite).ToArray(),
    favoriteSurveys.Select(t => t.Template).ToArray());
}
```

This method first instantiates a mock survey store and store locator objects to use in the test. The method then instantiates the view model object from the Tailspin.PhoneClient project to test, passing in the mock store locator object, along with other mock services. The method then executes the test on the view model instance and verifies that the favorite surveys in the view are the same as the ones in the underlying database.

For more information about testing the application see Appendix A, "Unit Testing Windows Phone Applications."

Displaying Data

The application displays data by binding elements of the view to properties in the view model. For example, the **Pivot** control on the SurveyListView page binds to the **SelectedPivotIndex** property in the **SurveyListViewModel** class.

The view can automatically update the values it displays in response to changes in the underlying view model if the view model implements the **INotifyPropertyChanged** interface. In the Tailspin mobile client, the abstract **ViewModel** class inherits from the **NotificationObject** class in the Prism Library. The **NotificationObject** class implements the **INotifyPropertyChanged** interface. With the exception of the question view models and the survey template view model, all the other view model classes in the Tailspin mobile client application inherit from the abstract **ViewModel** class. The application also uses the **ObservableCollection** class from the **System.Collections.ObjectModel** namespace that also implements the **INotifyPropertyChanged** interface.

*The view model implements the **INotifyPropertyChanged** interface indirectly through the **NotificationObject** class from the Prism Library.*

To learn more about Prism, see the Prism CodePlex site and Prism on MSDN.

Inside the Implementation

The following sections describe examples of data binding in the application. The first section describes a simple scenario on the AppSettings-View page, the next sections describe more complex examples using Pivot controls, and the last section describes how Tailspin addressed an issue with focus events on the phone.

Data Binding on the Settings Screen

Define your data bindings to the view model in the view's XAML.

The AppSettingsView page illustrates a simple scenario for binding a view to a view model. On this screen, the controls on the screen must display property values from the **AppSettingsViewModel** class, and set the property values in the view model when the user edits the settings values.

The following code example shows the **DataContext** attribute and some of the control definitions in the AppSettingsView.xaml file. Tailspin chose to use the **ToggleSwitch** control in place of a standard **CheckBox** control because it better conveys the idea of switching something on and off instead of selecting something. The **Toggle-Switch** control is part of the Microsoft Silverlight Windows Phone Toolkit available on the *Silverlight Toolkit* CodePlex site.

XAML

```
<phone:PhoneApplicationPage
  x:Class="TailSpin.PhoneClient.Views.AppSettingsView"
  ...
  DataContext="{Binding Source={StaticResource ViewModelLocator},
    Path=AppSettingsViewModel}"
  ...>
...
<shell:SystemTray.ProgressIndicator>
    <shell:ProgressIndicator IsIndeterminate="True"
      IsVisible="{Binding IsSyncing}"
      Text="{Binding ProgressText}"/>
</shell:SystemTray.ProgressIndicator>
...
<TextBox Height="Auto" HorizontalAlignment="Stretch"
  Margin="0,28,0,0" Name="textBoxUsername"
  VerticalAlignment="Top" Width="Auto"
  Text="{Binding UserName, Mode=TwoWay}" Padding="0"
  BorderThickness="3">
    <i:Interaction.Behaviors>
        <prism:UpdateTextBindingOnPropertyChanged/>
    </i:Interaction.Behaviors>
</TextBox>
```

```
...
<PasswordBox Height="Auto" HorizontalAlignment="Stretch"
  Margin="0,124,0,0" Name="passwordBoxPassword"
  VerticalAlignment="Top" Width="Auto"
  Password="{Binding Password, Mode=TwoWay}">
    <i:Interaction.Behaviors>
        <prism:UpdatePasswordBindingOnPropertyChanged/>
    </i:Interaction.Behaviors>
</PasswordBox>
...
<toolkit:ToggleSwitch Header="Subscribe to Push Notifications"
  Margin="0,202,0,0"
  IsChecked="{Binding SubscribeToPushNotifications, Mode=TwoWay}"
/>
...
```

The default binding mode value is **OneWay**, which is the setting used by the **ProgressIndicator** control. You need to change this to **TwoWay** if you want to send the changes back to the view model.

If a view wants to update its view model, the binding mode must be set to **TwoWay**. In order for the view model to notify the view of updates, the view model must implement the **INotifyPropertyChanged** interface. In the Tailspin client application, this interface is implemented by the **ViewModel** class from which all the view models inherit. A view model notifies a view of a changed property value by invoking the **RaisePropertyChanged** method. The following code example shows how the **AppSettingsViewModel** view model class notifies the view that it should display the in-progress indicator to the user.

C#

```
public bool IsSyncing
{
  get { return this.isSyncing; }
  set
  {
    this.isSyncing = value;
    this.RaisePropertyChanged(() => this.IsSyncing);
  }
}
```

The **RaisePropertyChanged** method uses an expression for compile-time verification. The Prism **PropertySupport** class performs the translation of a lambda expression to a property name.

The code for the AppSettingsView page illustrates a solution to a common issue in Silverlight for the Windows Phone platform. By default, the view does not notify the view model of property value changes until the control loses focus. For example, if the user enters a value in a control and then causes the control to lose focus, the view model will be updated. However, if the user enters a value in a control and then interacts with any **ApplicationBarButton** elements, the focus lost event will not be fired. For example, new content in the **textBoxUserName** control is lost if the user tries to save the settings before moving to another control. The **UpdateTextBindingOnPropertyChanged** behavior from the Prism Library ensures that the view always notifies the view model of any changes in the **TextBox** control as soon as they happen. The **UpdatePasswordBindingOnPropertyChanged** behavior does the same for the **PasswordBox** control. For more information, see the section, "Handling Focus Events," later in this chapter.

Data Binding and the Pivot Control

The application uses the **Pivot** control to allow the user to view different filtered lists of surveys. Figure 6 shows the components in the mobile client application that relate to the **Pivot** control as it's used on the SurveyListView page.

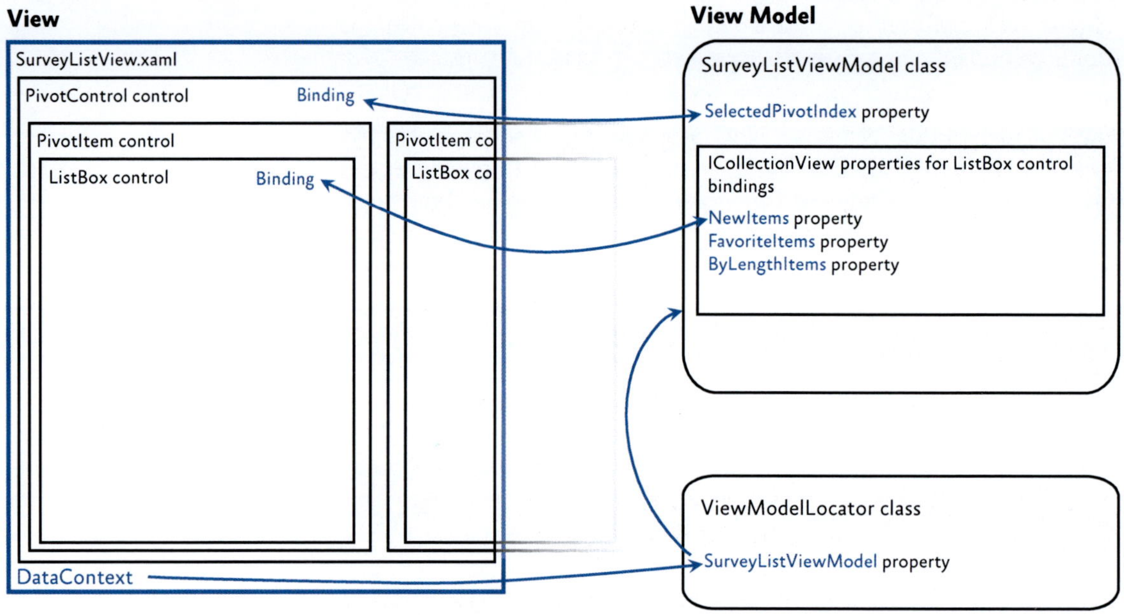

FIGURE 6
Using the Pivot control on the SurveyListView page

The following code example shows how the **Pivot** control on the SurveyListView page binds to the **SelectedPivotIndex** property of the **SurveyListViewModel** instance. This two-way binding determines which **PivotItem** control, and therefore which list of surveys, is displayed on the page. Remember, the **ViewModelLocator** class is responsible for locating the correct view model for a view.

```
XAML
<phoneControls:Pivot
  Title="TAILSPIN SURVEYS"
  Name="homePivot"
  SelectedIndex="{Binding SelectedPivotIndex, Mode=TwoWay}"
  Visibility="{Binding SettingAreConfigured,
    Converter={StaticResource VisibilityConverter}}">
  ...
</phoneControls:PivotControl>
```

The following code example shows the definition of the **PivotItem** control that holds a list of new surveys; it also shows how the **ListBox** control binds to the **NewItems** property in the view model.

```
XAML
<phoneControls:PivotItem Header="new">
  <Grid>
    <ContentControl Template="{StaticResource NoItemsTextBlock}"
      Visibility="{Binding NewItems.IsEmpty,
      Converter={StaticResource VisibilityConverter}}" />
    <ListBox ItemsSource="{Binding NewItems}"
      Style="{StaticResource SurveyTemplateItemStyle}"
      Visibility="{Binding NewItems.IsEmpty,
      Converter={StaticResource NegativeVisibilityConverter}}" >
    </ListBox>
  </Grid>
</phoneControls:PivotItem>
```

*In the list, the layout and formatting of each survey's information is handled by the **SurveyTemplate-ItemStyle** style and the **SurveyDataTemplate** data template in the Styles.xaml file.*

The **SurveyListViewModel** class uses **CollectionViewSource** objects to hold the list of surveys to display in the list on each **PivotItem** control. This allows the view model to notice and to react to changes in the item selected in the view, without needing to know about the view itself. The following code example shows how the **SurveyListViewModel** class defines the properties that the **ListBox** controls bind to.

```
C#
private CollectionViewSource newSurveysViewSource;
private CollectionViewSource byLengthViewSource;
private CollectionViewSource favoritesViewSource;
...
public ICollectionView NewItems
{
  get { return this.newSurveysViewSource.View; }
}

public ICollectionView FavoriteItems
{
  get { return this.favoritesViewSource.View; }
}

public ICollectionView ByLengthItems
{
  get { return this.byLengthViewSource.View; }
}
```

The following code example shows how the **BuildPivotDimensions** method populates the **CollectionViewSource** objects. In this example, to save space, only the code that populates the **newSurveysViewSource** property is shown.

```
C#
private void BuildPivotDimensions()
{
  ...
  this.ObservableSurveyTemplates =
    new ObservableCollection<SurveyTemplateViewModel>();
  List<SurveyTemplateViewModel> surveyTemplateViewModels =
    this.surveyStoreLocator.GetStore().GetSurveyTemplates()
    .Select(t => new SurveyTemplateViewModel(t,
      this.NavigationService,
      this.PhoneApplicationServiceFacade,
      this.shellTile,
      this.locationService)
      {
        CompletedAnswers = this.surveyStoreLocator.GetStore()
          .GetCurrentAnswerForTemplate(t) != null
          ? this.surveyStoreLocator.GetStore()
          .GetCurrentAnswerForTemplate(t).CompletedAnswers
          : 0,
```

```
        Completeness = this.surveyStoreLocator.GetStore()
          .GetCurrentAnswerForTemplate(t) != null
          ? this.surveyStoreLocator.GetStore()
          .GetCurrentAnswerForTemplate(t).CompletenessPercentage
          : 0,
        CanTakeSurvey = () => !IsSynchronizing
      }).ToList();
  surveyTemplateViewModels.ForEach(
    this.observableSurveyTemplates.Add);
  ...
  this.newSurveysViewSource = new CollectionViewSource
    { Source = this.observableSurveyTemplates };
  ...

  this.newSurveysViewSource.Filter +=
    (o, e) => e.Accepted =
      ((SurveyTemplateViewModel)e.Item).Template.IsNew;
  ...

  this.newSurveysViewSource.View.CurrentChanged +=
    (o, e) => this.SelectedSurveyTemplate =
      (SurveyTemplateViewModel)this.newSurveysViewSource
      .View.CurrentItem;
  ...

  // Initialize the selected survey template.
  this.HandleCurrentSectionChanged();
  ...
}
```

This method creates an **ObservableCollection** collection of **SurveyTemplateViewModel** objects. Each **SurveyTemplateViewModel** object holds a complete definition of a survey, its questions, and its answers. The method then assigns this collection to the **Source** property of each **CollectionViewSource** object. Next, the method applies a filter or a sort to each **CollectionViewSource** object so that it displays the correct set of surveys. The method then attaches an event handler to the **CurrentChanged** event of the view in each **CollectionViewSource** object so that the **SelectedSurveyTemplate** property of the **SurveyListViewModel** object is updated correctly. Finally, the method calls the **HandleCurrentSectionChanged** method that causes the view model to set the selected survey to the value of the **SelectedSurveyTemplate** property.

The **ObservableCollection** class provides notifications when the collection is modified. This means the view automatically updates through the bindings when the data changes.

The application also uses the **Pivot** control to display survey questions. This enables the user to scroll left and right through the questions as if the questions are all on a single large page of which the phone's screen shows a small portion.

Figure 7 shows how the **Pivot** control's binding works on the TakeSurveyView page.

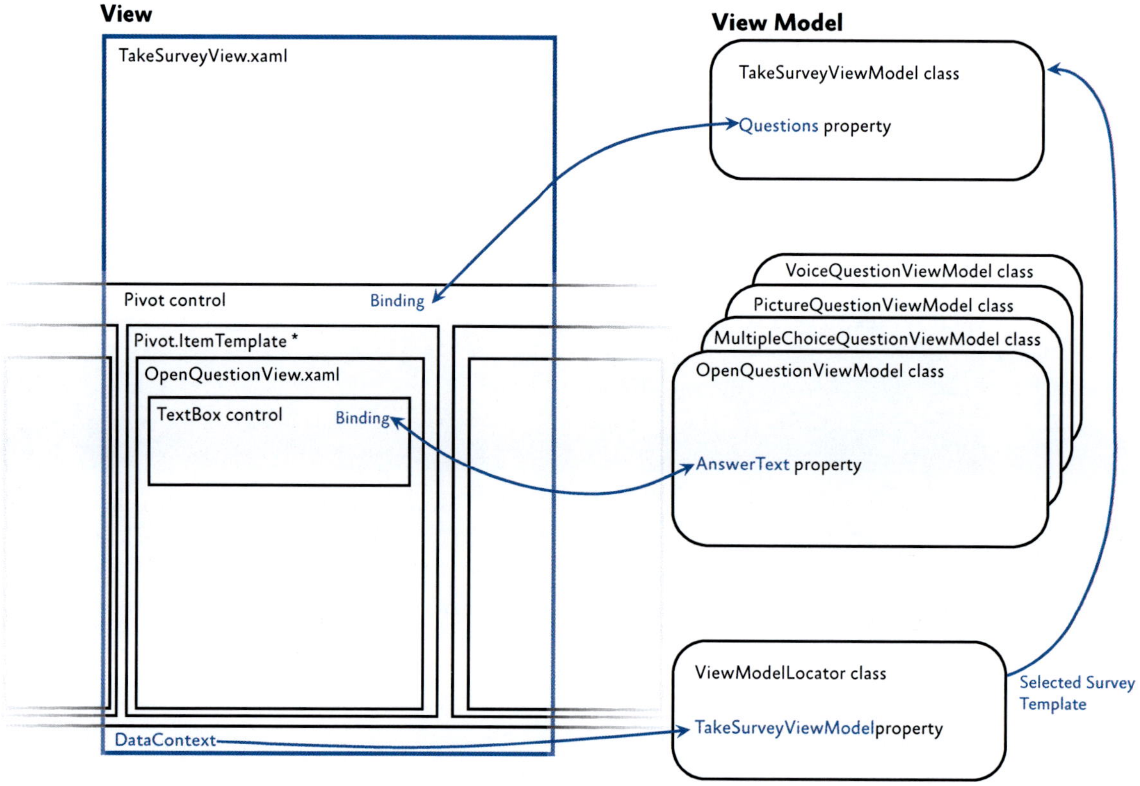

* The Pivot.ItemTemplate uses the DataTemplateSelector content selector class from the Prism Library to display the correct view based on the question type. The possible views are: FiveStarsQuestionView, MultipleChoiceQuestionView, OpenQuestionView, PictureQuestionView, and VoiceQuestionView.

FIGURE 7
Using the Pivot Control on the TakeSurveyView page

As in the previous examples, the view uses the **ViewModelLocator** class to create the view model. The following code example shows how the **ViewModelLocator** object instantiates a **TakeSurveyViewModel** object.

C#
```
public TakeSurveyViewModel TakeSurveyViewModel
{
  get
  {
    var vm = new TakeSurveyViewModel(
    (
      this.containerLocator.Container
        .Resolve<INavigationService>(),
      this.containerLocator.Container
        .Resolve<IPhoneApplicationServiceFacade>(),
      this.containerLocator.Container
        .Resolve<ILocationService>(),
      this.containerLocator.Container
        .Resolve<ISurveyStoreLocator>(),
      this.containerLocator.Container
        .Resolve<IShellTile>()
    );

    return vm;
  }
}
```

The **Pivot** control binds to the **Questions** property of the **TakeSurveyViewModel** class and a **Pivot.ItemTemplate** template controls the display of each question in the survey. However, it's necessary to display different question types using different layouts. The following code example from the TakeSurveyView.xaml file shows how the data binding and view layout is defined for the **Pivot** control using the **DataTemplate-Selector** content selector class from the Prism Library.

XAML
```
<phoneControls:Pivot
  SelectionChanged="PivotSelectionChanged"
  Loaded="ControlLoaded"
  VerticalAlignment="Top"
  Name="questionsPivot" Margin="0,0,0,0"
  ItemsSource="{Binding Questions}">
  <phoneControls:Pivot.ItemTemplate>
    <DataTemplate>
      <ScrollViewer>
        <prismViewModel:DataTemplateSelector Content="{Binding}"
          Grid.Row="0" VerticalAlignment="Top"
          HorizontalContentAlignment="Left" IsTabStop="False">
          <prismViewModel:DataTemplateSelector.Resources>
```

```
          <DataTemplate x:Key="OpenQuestionViewModel">
            <Views:OpenQuestionView DataContext="{Binding}"/>
          </DataTemplate>
          <DataTemplate x:Key="MultipleChoiceQuestionViewModel">
            <Views:MultipleChoiceQuestionView DataContext="{Binding}"/>
          </DataTemplate>
          ...
        </prismViewModel:DataTemplateSelector.Resources>
      </prismViewModel:DataTemplateSelector>
    </ScrollViewer>
  </DataTemplate>
</phoneControls:Pivot.ItemTemplate>
<phoneControls:Pivot.HeaderTemplate>
  <DataTemplate>
    <TextBlock Text="{Binding Title}"/>
  </DataTemplate>
</phoneControls:Pivot.HeaderTemplate>
</phoneControls:Pivot>
```

*You'll notice that the XAML declares handlers for the **SelectionChanged** and **Loaded** events in the code-behind. For an explanation of why the developers at Tailspin used code-behind here, see the section, "Handling Activation and Deactivation," in Chapter 3, "Using Services on the Phone." The code-behind also contains a workaround method to trim long titles that don't always display correctly when the user scrolls in the **Pivot** control.*

Each question view on the **Pivot** control binds to a view model object in the list of questions held in the **Questions** property of the **TakeSurveyViewModel** class. For example, an **OpenQuestionView** view binds to an **OpenQuestionViewModel** object, and a **VoiceQuestionView** view binds to a **VoiceQuestionViewModel** object. The following code example shows how the **TakeSurveyViewModel** class builds this list of question view models.

C#

```
public IList<QuestionViewModel> Questions { get; set; }
...
public void Initialize(ISurveyStoreLocator surveyStoreLocator)
{
    ...
    this.Questions = this.SurveyAnswer.Answers.Select(
      a => Maps[a.QuestionType].Invoke(a)).ToArray();
}
```

The following code sample shows the definition of the **Maps** property in the **TakeSurveyViewModel**. The **Maps** property maps question types to view models.

```
C#
private static readonly
  Dictionary<QuestionType, Func<QuestionAnswer,
  QuestionViewModel>> Maps =
  new Dictionary<QuestionType, Func<QuestionAnswer,
  QuestionViewModel>>()
  {
    { QuestionType.SimpleText,
      a => new OpenQuestionViewModel(a) },
    { QuestionType.MultipleChoice,
      a => new MultipleChoiceQuestionViewModel(a) },
    ...
  };
```

Handling Focus Events

The file OpenQuestionView.xaml defines the UI for entering survey results. Tailspin found that when the user clicked the **Save** button, the last answer wasn't saved by the view model. This is because **ApplicationBarIconButton** control is not a **FrameworkElement** control, so it cannot get the focus. As a result, the lost focus event on the text box wasn't being raised; as a result, the bindings didn't tell the view model about the new field contents.

To solve this problem, the developers at Tailspin used a behavior named **UpdateTextBindingOnPropertyChanged** from the Prism Library. This behavior ensures that the view notifies the view model whenever the user changes the text in the **TextBox** control, not just when the control loses focus. The following code example shows how this behavior is defined in OpenQuestionView.xaml.

```
XAML
...
xmlns:prism=
"clr-namespace:Microsoft.Practices.Prism.Interactivity;
assembly=Microsoft.Practices.Prism.Interactivity"
...
<TextBox Text="{Binding AnswerText, Mode=TwoWay}" Height="100">
   <Interactivity:Interaction.Behaviors>
      <prism:UpdateTextBindingOnPropertyChanged/>
   </Interactivity:Interaction.Behaviors>
</TextBox>
```

Commands

In a Silverlight application, you can invoke some action in response to a user action (such as a button click) by creating an event handler in the code-behind class. However, in the MVVM pattern, the responsibility for implementing the action lies with the view model, and you should avoid placing code in the code-behind classes. Therefore, you should connect the control to a method in the view model using a command binding.

To keep the responsibilities of the view and view model separate, you should try to avoid placing code in the code-behind files of your views when you are implementing the MVVM pattern.

In Silverlight 4, the **ButtonBase** class and the **Hyperlink** class both support **Command** and **CommandParameter** properties. The **Command** property can reference an ICommand implementation that comes from a view model data source, through a binding. The command is then interpreted at runtime by the Silverlight input system. For more information, see *"ButtonBase.Command Property"* on MSDN and *"Hyperlink.Command Property"* on MSDN.

However, because the **ApplicationBarIconButton** class is not a control, Tailspin uses the **ApplicationBarButtonCommand** behavior from the Prism Library to bind the click event of the **ApplicationBar-ButtonCommand** to the execution of a command.

For more information about the **ICommand** interface, see *"ICommand Interface"* on MSDN.

Inside the Implementation

Windows Phone controls that derive from ***ButtonBase*** *or* ***Hyperlink*** *support binding to* ***ICommand*** *instances.*

The developers at Tailspin use bindings to associate actions in the UI with commands in the view model. However, you cannot use the **InvokeCommandAction** Expression Blend behavior with the **ApplicationBarIconButton** or **ApplicationBarMenuItems** controls because they cannot have dependency properties. Tailspin uses a custom behavior to connect commands to the view model.

The following code example from the SurveyListView page shows how the **Synchronize** button, the **Settings** button, and the **Filter** button on the application bar are associated with commands. In addition, it shows how the **About** menu item on the application bar is associated with an event handler for the **Click** event.

XAML

```
<phoneNavigation:PhoneApplicationPage.ApplicationBar>
  <shell:ApplicationBar IsVisible="True">
    <shell:ApplicationBar.MenuItems>
      <shell:ApplicationBarMenuItem
        Text="about"
        Click="AboutMenuItem_Click"/>
      </shell:ApplicationBar.MenuItems>
    <shell:ApplicationBarIconButton Text="Sync" IconUri="..." />
    <shell:ApplicationBarIconButton Text="Settings"
      IconUri="..." />
```

```
    <shell:ApplicationBarIconButton Text="Filters" IconUri="..."/>
  </shell:ApplicationBar>
</phoneNavigation:PhoneApplicationPage.ApplicationBar>
  ...
<Custom:Interaction.Behaviors>
  <prismInteractivity:ApplicationBarButtonCommand
    ButtonText="Sync"
    CommandBinding="{Binding StartSyncCommand}"/>
  <prismInteractivity:ApplicationBarButtonCommand
    ButtonText="Settings"
    CommandBinding="{Binding AppSettingsCommand}" />
  <prismInteractivity:ApplicationBarButtonCommand
    ButtonText="Filters"
    CommandBinding="{Binding FiltersCommand}" />
</Custom:Interaction.Behaviors>
```

The attributes attached to the **ApplicationBarIconButton** controls (**Text** and **IconUri**) only affect their appearance. The **ApplicationBarButtonCommand** elements handle the connection to the commands; they identify the control to associate with the command through the **ButtonText** attribute and the command through the **CommandBinding** attribute.

The **ApplicationBarButtonCommand** class from the Prism Library implements the custom behavior that links a button click in the UI to the **StartSyncCommand**, **AppSettingsCommand**, and **FiltersCommand** properties in the **SurveyListViewModel** class.

The **StartSyncCommand** property uses an instance of the **DelegateCommand** class that implements the **ICommand** interface. The following code example from the **SurveyListViewModel** class shows the definition of the **StartSyncCommand** property.

C#

```
public DelegateCommand StartSyncCommand { get; set; }
...
public SurveyListViewModel(...) : base(...)
{
  ...
  this.StartSyncCommand = new DelegateCommand(
    this.StartSync,
    () => !this.IsSynchronizing &&
          !this.SettingAreNotConfigured);
    ...
}
```

For more information about the ***DelegateCommand*** *class from Prism, see Chapter 9 of the Prism documentation, "Communicating Between Loosely Coupled Components" on MSDN.*

The following code example from the **SurveyListViewModel** class shows the implementation of the **StartSync** method and **SyncCompleted** method. The **StartSync** method runs the synchronization process asynchronously by invoking the **StartSynchronization** method. It also uses the **ObserveOnDispatcher** method from the Reactive Extensions (Rx) for .NET. For more information about how Tailspin uses Rx, see Chapter 3, "Using Services on the Phone."

C#

```
public void StartSync()
{
  if (this.IsSynchronizing)
  {
    return;
  }

  this.IsSynchronizing = true;
  this.synchronizationService
    .StartSynchronization()
    .ObserveOnDispatcher()
    .Subscribe(this.SyncCompleted);
}

private void SyncCompleted(
  IEnumerable<TaskCompletedSummary> taskSummaries)
{
  ...
  this.BuildPivotDimensions();
  this.IsSynchronizing = false;
  this.UpdateCommandsForSync();
}
```

The **SyncCompleted** method also updates the UI to show the new list of surveys following the synchronization process; it also controls the progress indicator in the UI by setting the **IsSynchronizing** property. The **UpdateCommandsForSync** method disables the **Synchronize** button in the UI while the synchronization is running.

Handling Navigation Requests

In addition to invoking commands from the view, the Tailspin mobile client also triggers navigation requests from the view. These requests could be to navigate to a particular view or navigate back to the previous view. In some scenarios, for example if the application needs to navigate to a new view when a command completes, the view model will need to send a message to the view. In other scenarios, you might want to trigger the navigation request directly from the view without involving the view model directly. When you're using the MVVM pattern, you want to be able to do all this without using any code-behind in the view, and without introducing any dependency on the view implementation in the view model classes.

Inside the Implementation

The following code example from the FilterSettingsView.xaml file shows how navigation is initiated in the sample application.

XAML

```
<i:Interaction.Behaviors>
  <prismInteractivity:ApplicationBarButtonCommand
    ButtonText="Save" CommandBinding="{Binding SaveCommand}"/>
  <prismInteractivity:ApplicationBarButtonCommand
    ButtonText="Cancel" CommandBinding="{Binding CancelCommand}" />
</i:Interaction.Behaviors>
```

In both cases, commands are invoked in the view model. The code that implements each command causes the application to navigate back to the previous view if the command succeeds, so the navigation is initiated from the view model. The following code example from the **FilterSettings-ViewModel** class illustrates this.

C#

```
public DelegateCommand SaveCommand { get; set; }
public DelegateCommand CancelCommand { get; set; }
...
public FilterSettingsViewModel(...)
{
  ...
  this.SaveCommand =
    new DelegateCommand(this.Submit, () => !this.CanSubmit);
  this.CancelCommand =
    new DelegateCommand(this.Cancel, () => true);
  ...
}

public bool CanSubmit
{
  get { return this.canSubmit; }
  set
  {
    if (!value.Equals(this.canSubmit))
    {
      this.canSubmit = value;
      this.RaisePropertyChanged(() => this.CanSubmit);
      this.SaveCommand.RaiseCanExecuteChanged();
    }
  }
}
...
public void Submit()
```

```
{
  ...
  if (this.NavigationService.CanGoBack) this.NavigationService.GoBack();
  ...
}

public void Cancel()
{
  if (this.NavigationService.CanGoBack) this.NavigationService.GoBack();
}

public override void OnPageDeactivation(bool isIntentionalNavigation)
{
  base.OnPageDeactivation(isIntentionalNavigation);

  if (isIntentionalNavigation)
  {
    this.Dispose();
    return;
  }
  ...
}
...
```

The **OnPageDeactivation** method is called by the **PhoneApplicationFrame Navigating** event when the page is intentionally or unintentionally navigated away from. The **isIntentionalNavigation** parameter indicates whether the current application is both the origin and destination of the navigation. Therefore, when navigating to another page in the application, the **Dispose** method of the base **ViewModel** class will be called in order to dispose of the instance of the **FilterSettingsViewModel**, provided that the navigation is intentional. For more information about the **PhoneApplicationFrame Navigating** event, see the section, "Handling Activation and Deactivation" in Chapter 3, "Using Service on the Phone."

Navigation is performed using the **ApplicationFrameNavigationService** class, from the TailSpin.Phone-Client.Adapters project, which is shown in the following code example.

```
C#
public class ApplicationFrameNavigationService :
  INavigationService
{
  private readonly PhoneApplicationFrame frame;
  private Dictionary<string, bool> tombstonedPages;

  public ApplicationFrameNavigationService(
    PhoneApplicationFrame frame)
```

```
{
  this.frame = frame;
  this.frame.Navigated += frame_Navigated;
  this.frame.Navigating += frame_Navigating;
  this.frame.Obscured += frame_Obscured;
  this.RecoveredFromTombstoning = false;
}

public bool CanGoBack
{
  get { return this.frame.CanGoBack; }
}

public bool RecoveredFromTombstoning { get; set; }

public bool DoesPageNeedtoRecoverFromTombstoning(Uri pageUri)
{
  if (!RecoveredFromTombstoning) return false;

  if (tombstonedPages == null)
  {
    tombstonedPages = new Dictionary<string, bool>();
    tombstonedPages.Add(pageUri.ToString(), true);
    foreach (var journalEntry in frame.BackStack)
    {
      tombstonedPages.Add(journalEntry.Source.ToString(), true);
    }
    return true;
  }

  if (tombstonedPages.ContainsKey(pageUri.ToString()))
  {
    return tombstonedPages[pageUri.ToString()];
  }
  return false;
}

public void UpdateTombstonedPageTracking(Uri pageUri)
{
  tombstonedPages[pageUri.ToString()] = false;
}
```

```
  public bool Navigate(Uri source)
  {
    return this.frame.Navigate(source);
  }

  public void GoBack()
  {
    this.frame.GoBack();
  }

  public event NavigatedEventHandler Navigated;

  void frame_Navigated(object sender, NavigationEventArgs e)
  {
    var handler = this.Navigated;
    if (handler != null)
    {
        handler(sender, e);
    }
  }

  public event NavigatingCancelEventHandler Navigating;

  void frame_Navigating(object sender, NavigatingCancelEventArgs e)
  {
    var handler = this.Navigating;
    if (handler != null)
    {
        handler(sender, e);
    }
  }

  public event EventHandler<ObscuredEventArgs> Obscured;

  void frame_Obscured(object sender, ObscuredEventArgs e)
  {
    var handler = this.Obscured;
    if (handler != null)
    {
        handler(sender, e);
    }
  }
}
```

This class, which implements Tailspin's **INavigationService** interface, uses the phone's **PhoneApplicationFrame** instance to perform the navigation request for the application.

A view model can invoke the **Navigate** method on the **ApplicationFrameNavigationService** object to cause the application to navigate to a particular view in the application or the **GoBack** method to return to the previous view.

The **ViewModel** base class maintains the **INavigationService** instance for all the view models, and the Funq dependency injection container is responsible for initially creating the **ApplicationFrameNavigationService** object that implements this interface.

Using the **PhoneApplicationFrame** instance ensures that the phone maintains the correct navigation stack for the application so that navigating backward works the way users expect.

To avoid any code-behind in the view when the view initiates the navigation, the developers at Tailspin use an interaction behavior from the Prism Library. The following code example shows how the **Cancel** button is declared in the FilterSettingsView.xaml file.

XAML
```
<phone:PhoneApplicationPage.ApplicationBar>
  <shell:ApplicationBar ...>
    ...
    <shell:ApplicationBarIconButton Text="Cancel" IconUri="..."/>
  </shell:ApplicationBar>
</phone:PhoneApplicationPage.ApplicationBar>
...
<i:Interaction.Behaviors>
  ...
  <prismInteractivity:ApplicationBarButtonCommand
    ButtonText="Cancel"
    CommandBinding="{Binding CancelCommand}" />
</i:Interaction.Behaviors>
```

By using a behavior, Tailspin avoids having any code in the view to handle navigation requests. This general pattern is how the view model can handle requests from the view without using "classic" events that require handlers in code-behind.

User Interface Notifications

The Tailspin Surveys mobile client application performs some tasks asynchronously; one example is the potentially time-consuming synchronization with the Tailspin Surveys web service. Asynchronous tasks must often inform the user of the outcome of the task or provide status information while they're running. It's important to consider the usability of the application when you're deciding on the appropriate way to notify users of significant events or to provide status information. The developers at Tailspin were concerned they would either flood the user with alerts or have the user miss an important piece of information.

For the Tailspin Surveys mobile client application, the developers identified two categories of notification, informational/warning notifications and error notifications.

Tailspin implemented a custom toast notification system.

Informational/Warning Notifications

Informational or warning notifications should not be disruptive, so the user should see the message but not be interrupted in their current task. The user does not need to perform any action in response to this type of message; the Tailspin mobile client application uses this type of notification to inform the user when a synchronization completes successfully, for example. Tailspin uses a custom toast notification for these messages because the application does not have access to the Windows Phone toast notification system.

Error Notifications

Error notifications should be disruptive because the message is informing the user that some expected action of the application will not happen. The notification should inform the user of the actions they need to take to resolve the problem; for example, to retry the synchronization operation again if it fails for some reason. Tailspin uses message boxes for this type of message.

Inside the Implementation

This example shows how Tailspin implements custom toast notifications and error notifications on the SurveyListView page to inform users when the synchronization process finishes or fails. In the sample application, many of the notifications and error messages are not intended for the imaginary user of the sample; instead, they are there to help you, the developer understand what the application is doing as you explore its functionality. You should follow the guidance published in the *User Experience Design Guidelines for Windows Phone* when you design the user notification system for your application.

The following code example shows the relevant declarations in the SurveyListView.xaml file.

XAML

```
<Custom:EventTrigger
  SourceObject="{Binding SubmitNotificationInteractionRequest}"
  EventName="Raised">
    <prismInteractionRequest:ToastPopupAction PopupElementName=
      "SynchronizationToast" />
</Custom:EventTrigger>

<Custom:EventTrigger
  SourceObject="{Binding SubmitErrorInteractionRequest}"
  EventName="Raised">
    <prismInteractionRequest:MessageBoxAction />
</Custom:EventTrigger>
...
```

```
<Popup x:Name="SynchronizationToast" DataContext="">
  <Grid x:Name="grid" Background="{StaticResource PhoneAccentBrush}"
    VerticalAlignment="Bottom" Width="480">
    <TextBlock Text="{Binding Title}"
      HorizontalAlignment="Stretch"
      Foreground="{StaticResource PhoneForegroundBrush}"
      TextWrapping="Wrap" Margin="14,5,14,5">
      <Custom:Interaction.Behaviors>
        <pag:PopupHideOnLeftMouseUp/>
      </Custom:Interaction.Behaviors>
    </TextBlock>
  </Grid>
</Popup>
```

The view model uses the **SubmitNotificationInteractionRequest** binding to trigger the toast notification and the **SubmitErrorInteractionRequest** binding to trigger the error message notification. The following code example shows how the **SurveyListViewModel** displays a toast notification when the synchronization process completes successfully and an error message when it fails.

```
C#
private readonly InteractionRequest
  <InteractionRequest.Notification> submitErrorInteractionRequest;
private readonly InteractionRequest
  <InteractionRequest.Notification> submitNotificationInteractionRequest;
...
public IInteractionRequest SubmitErrorInteractionRequest
{
  get { return this.submitErrorInteractionRequest; }
}

public IInteractionRequest SubmitNotificationInteractionRequest
{
  get { return this.submitNotificationInteractionRequest; }
}
...
private void SyncCompleted(
  IEnumerable<TaskCompletedSummary> taskSummaries)
{
  ...
    if (taskSummaries.Any(
      t => t.Result != TaskSummaryResult.Success))
    {
      this.submitErrorInteractionRequest.Raise(
        new InteractionRequest.Notification
        {
```

```
          Title = "Synchronization error",
          Content = stringBuilder.ToString()
        },
        n => { });
    }
    else
    {
      this.submitNotificationInteractionRequest.Raise(
        new InteractionRequest.Notification
        {
          Title = stringBuilder.ToString(),
          Content = null
        },
        n => { });
  }
  ...
}
```

*This solution uses the **InteractionRequest** and **Notification** classes and two trigger actions, **MessageBoxAction** and **ToastPopupAction**, from the Prism Library.*

Accessing Services

The MVVM pattern identifies three major components: the view, the view model, and the model. This chapter describes the UI of the Tailspin Surveys mobile client application and the way that Tailspin uses the MVVM pattern. The Tailspin mobile client application also includes a number of services. These services can be invoked from the view, view model, or model components and include services that do the following:

- Manage the settings and surveys stores that handle data persistence for the application.
- Save and load the application's state when it's activated and deactivated.
- Synchronize survey data between the client application and the Tailspin Surveys web application.
- Notify the application when new surveys are available.
- Encode audio data, and support application navigation and other infrastructure services.

These services are discussed in the following chapters. Chapter 3, "Using Services on the Phone," describes how the application uses services offered by the Windows Phone platform, such as local data persistence services and geo-location services. Chapter 4, "Connecting with Services," describes how the mobile client application accesses services over the network.

Conclusion

This chapter described how the developers at Tailspin built the UI components of the application, and how and why the MVVM pattern was implemented. The next chapter will describe how the developers at Tailspin implemented the model elements from the MVVM pattern in the mobile client application, and how the application leverages services offered by the Windows Phone platform, such as isolated storage, background agents, and location services.

Questions

1. Which of the following are good reasons to use the MVVM pattern for your Windows Phone application?
   a. It improves the testability of your application.
   b. It facilitates porting of the application to another platform, such as the desktop.
   c. It helps to make it possible for designers and developers to work in parallel.
   d. It may help you avoid risky changes to existing model classes.
2. Which of the following are good reasons not to use the MVVM pattern for your Windows Phone application?
   a. You have a very tight deadline to release the application.
   b. Your application is relatively simple with only two screens and no complex logic to implement.
   c. Windows Phone controls are not ideally suited to the MVVM pattern.
   d. It's unlikely that your application will be used for more than six months before it is completely replaced.
3. Which of the following are correct about tombstoning?
   a. Tombstoned applications have been terminated.
   b. Tombstoned applications remain intact in memory.
   c. Information about a tombstoned application's navigation state and state dictionaries are preserved for when the application is relaunched.
   d. A device will maintain tombstoning information for up to five applications at once.
4. Which of the following describe the role of the view model locator?
   a. The view model locator configures bindings in the MVVM pattern.
   b. In the Tailspin mobile client, the view model locator is responsible for instantiating view-model objects.
   c. The view model locator connects views to view models.
   d. Data template relations offer an alternative approach to a view model locator.
5. Where does the Back button take you?
   a. To the previous view in the navigation stack.
   b. It depends on what the code in the view model does.
   c. If the current view is the last one in the navigation stack, you leave the application.
   d. If your application is on the top of the phone's application stack, it takes you back to your application.

6. Why should you not use code-behind when you're using the MVVM pattern?
   a. The view model locator always intercepts the events, so code-behind code never executes.
   b. The MVVM pattern enforces a separation of responsibilities between the view and the view model. UI logic belongs in the view model.
   c. If you are using the MVVM pattern, other developers will expect to see your code in the view model classes and not in the code-behind.
   d. Code-behind has a negative effect on view performance.

More Information

For more information about designing a Windows Phone UI, see "*Themes for Windows Phone*" on MSDN.

For more information about the **Pivot** control, see "*Pivot Control for Windows Phone*" on MSDN.

For more information about navigation on the Windows Phone platform, see "*Frame and Page Navigation for Windows Phone*" on MSDN.

For more information about Prism and MVVM see the *Prism* CodePlex site and "*Prism*" on MSDN.

These and all links in this book are accessible from the book's online bibliography. You can find the bibliography on MSDN at: ***http://msdn.microsoft.com/en-us/library/gg490786.aspx***.

3 Using Services on the Phone

This chapter describes how the Tailspin mobile client uses services offered by the Windows® Phone platform. The chapter begins by describing the various model classes used in the application to represent data within the application. The view model classes described in Chapter 2, "Building the Mobile Client," make extensive use of these model classes as they manage the data displayed and created in the user interface (UI). This chapter focuses on how the mobile client application uses the isolated storage service on the phone to persist this data, and how the application can populate these model classes with previously saved data. The chapter also discusses issues that relate to data security on the phone.

When the Windows Phone operating system deactivates an application, it's the application's responsibility to persist enough state information to be able to restore the state of the application if and when it's reactivated by the operating system. This chapter describes how the Tailspin mobile client uses the services offered by the phone to support this behavior.

The mobile client application can also send and receive data from the Tailspin cloud service, and this chapter describes how the application synchronizes data between the phone and the cloud. The focus of this chapter is the way that the mobile client application can use services on the phone to help it run the synchronization tasks both in the background and in the foreground.

The chapter also describes how the mobile client application uses other services to pin tiles to Start, and to acquire geo-location data, image data, and audio data.

The Model Classes

Chapter 2, "Building the Mobile Client," described the Model-View-ViewModel (MVVM) pattern adopted by the developers at Tailspin. This chapter relates to the model elements of this pattern, contained in the TailSpin.PhoneServices project, that represent the domain entities used in the application. Figure 1 shows the key model classes and the relationships between them.

Model classes represent domain entities in the Model-View-ViewModel pattern.

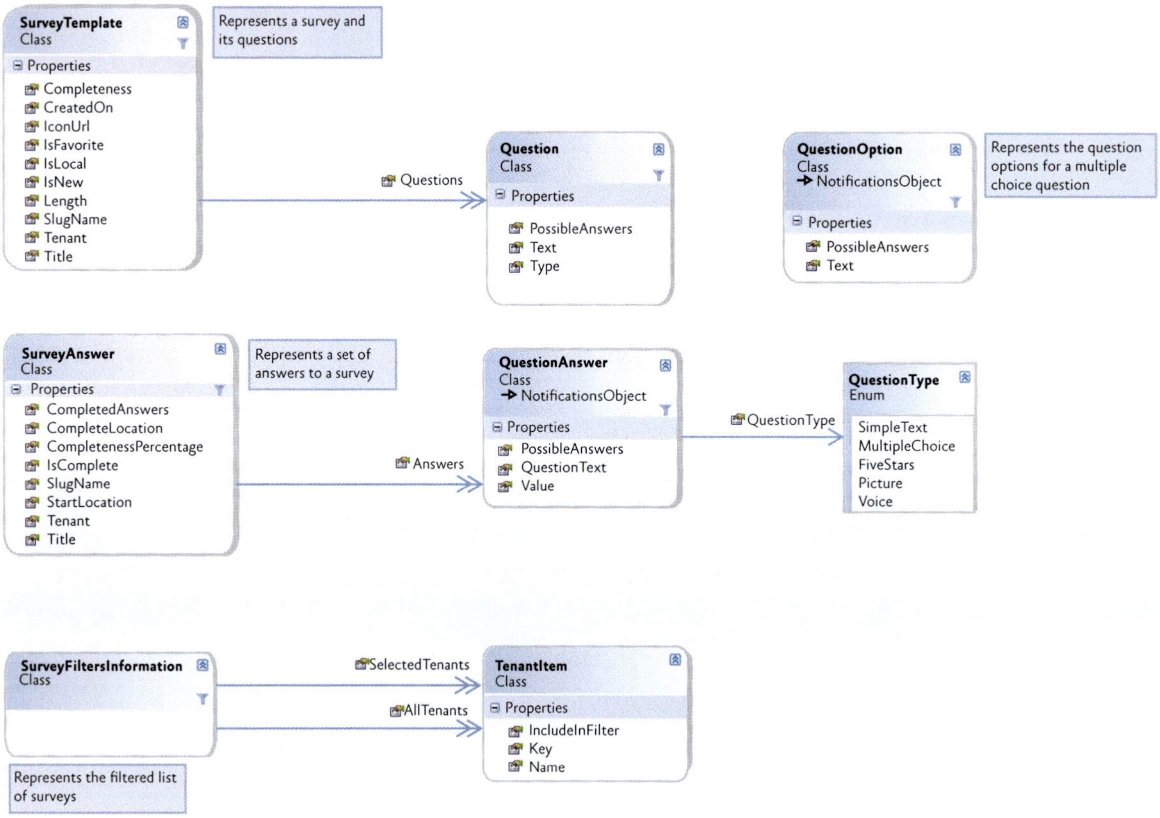

FIGURE 1
The model classes in the Tailspin mobile client application

Using Isolated Storage on the Phone

Isolated storage is the only mechanism available to persist application data on the Windows Phone platform. When an application uses a local database, it resides in the application's isolated storage container.

The mobile client application must be able to operate when there is no available network connection. Therefore, it must be able to store survey definitions and survey answers locally on the device, together with any other data or settings that the application requires to operate.

The application must behave as a "good citizen" on the device. It is not possible to access another application's isolated storage or access the file system directly on the Windows Phone platform, so the application cannot compromise another application by reading or modifying its data. However, there is a limited amount of storage available on the phone, so Tailspin tries to minimize the amount of data that the mobile client application stores locally and be proactive about removing unused data.

The Tailspin mobile client application must adopt a robust storage solution and minimize the risk of losing any stored data.

Overview of the Solution

The Windows Phone platform offers isolated storage functionality. Isolated storage provides a dictionary-based model for storing application settings, enables the application to create folders and files that are private to the application, and can store relational data in a local database by using LINQ to SQL.

Figure 2 shows how each application on a Windows Phone device only has access to its own private storage on the device, and cannot access the storage belonging to other applications.

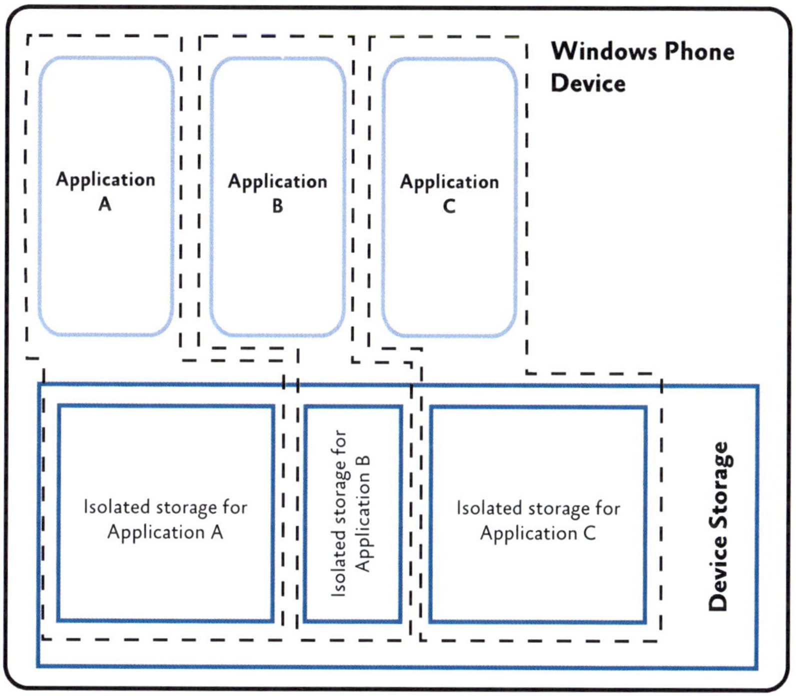

Figure 2
Isolated storage on the Windows Phone platform

To be a "good citizen" on the phone, your application should minimize the amount of isolated storage it uses. There's nothing on the phone that enforces any quotas, it's your responsibility.

It is the application's responsibility to manage the amount of storage it uses by deleting temporary or unused data because the operating system does not enforce any quota limits. Tailspin removes completed surveys from isolated storage after they are synchronized to the Tailspin Surveys service.

Isolated storage provides one level of data security—isolation from other applications on the device—but you should consider whether this is adequate for the security needs of your application.

Security

Tailspin does not encrypt the survey data that the mobile client application stores in isolated storage because it does not consider that data to be confidential. However, Tailspin does encrypt user passwords before storing them in isolated storage.

> *The sample application stores user passwords as encrypted text in isolated storage.*

There are two scenarios to think about when you are implementing security for data stored on the phone. The first is whether other applications on the phone could potentially access your application's data and then transmit it to someone else. The isolated storage model used on Windows Phone that limits applications to their own storage makes this a very unlikely scenario. However, security best practices suggest that you should guard against even unlikely scenarios, so you may want to consider encrypting the data in your application's isolated storage.

The second scenario to consider is what happens if an unauthorized user gains access to the device. If you want to protect your data in this scenario, you must encrypt the data while it is stored on the device.

The Windows Phone API provides access to the Data Protection API (DPAPI). DPAPI generates and stores a cryptographic key by using the user and phone credentials to encrypt and decrypt data. The **ProtectedData** class provides access to DPAPI through **Protect** and **Unprotect** methods. On a Windows Phone device, every application gets its own decryption key, which is created when an application is run for the first time. Calls to **Protect** and **Unprotect** will implicitly use the decryption key and make sure that all the data remains private to the application. In addition, the decryption key persists across application updates. For more information, see *"Security for Windows Phone,"* on the MSDN® developer program website.

Storage Format

Tailspin stores the application's settings as key/value pairs and uses serializable model classes to store the survey data as files in isolated storage. The latest release of the Windows Phone platform includes the ability to store data in a local database, and the developers at Tailspin are considering this option for the future. They have implemented the storage service in the application in a way that makes it easy to replace the storage classes with alternative implementations if they decide to use a different storage mechanism in the future. The current implementation also makes it easy to test the storage functionality in the application.

Inside the Implementation

Now is a good time to walk through the code that implements isolated storage in the Tailspin mobile client application in more detail. As you go through this section, you may want to download the *Windows Phone Tailspin Surveys* from the Microsoft Download Center.

You can find the code that implements isolated storage access in the Tailspin mobile client application in the Services/Stores folder in the TailSpin.PhoneServices project.

Application Settings

The user enters application settings data on the AppSettingsView page in the Surveys application. The **ISettingsStore** interface defines the data items that the application saves as settings. The following code example shows this interface.

C#

```
public interface ISettingsStore
{
  string Password { get; set; }
  string UserName { get; set; }
  bool SubscribeToPushNotifications { get; set; }
  bool LocationServiceAllowed { get; set; }
  bool BackgroundTasksAllowed { get; set; }
  event EventHandler UserChanged;
}
```

The following code example shows how the application implements this interface to save the value of the **Password** property in the **ApplicationSettings** dictionary in the application's isolated storage.

C#

```
public class SettingsStore : ISettingsStore
{
  private readonly IProtectData protectDataAdapter;
  private const string PasswordSettingDefault = "";
  private const string PasswordSettingKeyName = "PasswordSetting";
  private readonly IsolatedStorageSettings isolatedStore;
  private UTF8Encoding encoding;
  ...

  public SettingsStore(IProtectData protectDataAdapter)
  {
    this.protectDataAdapter = protectDataAdapter;
    isolatedStore = IsolatedStorageSettings.ApplicationSettings;
    encoding = new UTF8Encoding();
  }

  public string Password
  {
    get
    {
```

```
      return PasswordByteArray.Length == 0 ? PasswordSettingDefault :
        encoding.GetString(PasswordByteArray, 0, PasswordByteArray.Length);
    }
    set
    {
      PasswordByteArray = encoding.GetBytes(value);
    }
  }

  private byte[] PasswordByteArray
  {
    get
    {
      byte[] encryptedValue = GetValueOrDefault(PasswordSettingKeyName,
        new byte[0]);
      if (encryptedValue.Length == 0)
        return new byte[0];
      return protectDataAdapter.Unprotect(encryptedValue, null);
    }
    set
    {
      byte[] encryptedValue = protectDataAdapter.Protect(value, null);
      AddOrUpdateValue(PasswordSettingKeyName, encryptedValue);
    }
  }

  ...

  private void AddOrUpdateValue(string key, object value)
  {
    bool valueChanged = false;

    try
    {
      // If the new value is different, set the new value.
      if (this.isolatedStore[key] != value)
      {
        this.isolatedStore[key] = value;
        valueChanged = true;
      }
    }
    catch (KeyNotFoundException)
```

```
      {
        this.isolatedStore.Add(key, value);
        valueChanged = true;
      }
      catch (ArgumentException)
      {
        this.isolatedStore.Add(key, value);
        valueChanged = true;
      }

      if (valueChanged)
      {
        Save();
      }
    }

    private T GetValueOrDefault<T>(string key, T defaultValue)
    {
      T value;

      try
      {
        value = (T)this.isolatedStore[key];
      }
      catch (KeyNotFoundException)
      {
        value = defaultValue;
      }
      catch (ArgumentException)
      {
        value = defaultValue;
      }

      return value;
    }

    private void Save()
    {
      isolatedStore.Save();
    }
}
```

The **SettingsStore** constructor accepts a parameter of **IProtectData**, which specifies the class that will provide access to the encryption functionality.

The **IProtectData** interface, in the TailSpin.Phone.Adapters project, defines method signatures for encrypting and decrypting data. The following code example shows this interface.

```
C#
public interface IProtectData
{
  byte[] Protect(byte[] userData, byte[] optionalEntropy);
  byte[] Unprotect(byte[] encryptedData, byte[] optionalEntropy);
}
```

The **IProtectData** interface is implemented by the **ProtectDataAdapter** class, which adapts the **ProtectedData** class that provides access to the Data Protection API. The purpose of adapting the **ProtectedData** class with a class that implements **IProtectData** is to create a loosely coupled class that is testable. The following code example shows the class.

An adapter is a design pattern that translates an interface for a class into a compatible interface. The adapter translates calls to its interface into calls to the original interface. This approach enables the writing of loosely coupled code that is testable.

```
C#
public class ProtectDataAdapter : IProtectData
{
  public byte[] Protect(byte[] userData, byte[] optionalEntropy)
  {
    return ProtectedData.Protect(userData, optionalEntropy);
  }

  public byte[] Unprotect(byte[] encryptedData,
    byte[] optionalEntropy)
  {
    return ProtectedData.Unprotect(encryptedData,
      optionalEntropy);
  }
}
```

Tailspin uses JavaScript Object Notation (JSON) serialization to reduce CPU usage and storage space requirements.

Therefore, rather than calling the **ProtectedData** class methods directly, the **PasswordByteArray** property in the **SettingsStore** class calls the methods in the **ProtectDataAdapter** class to perform data encryption and decryption.

Survey Data

The application saves the local survey data in isolated storage as serialized **SurveyTemplate** and **SurveyAnswer** objects. The following code example shows the definition of the **SurveysList** object that uses the model classes **SurveyTemplate** and **SurveyAnswer**. The **SurveyTemplate** class is a model class that defines a survey and includes the question types, question text, possible answers, and survey metadata. The **SurveyAnswer** class is a model class that defines the responses collected by the surveyor. For more information about these model classes, see the section, "The Model Classes," earlier in this chapter.

C#
```
public class SurveysList
{
  public SurveysList()
  {
    this.LastSyncDate = string.Empty;
  }

  public List<SurveyTemplate> Templates { get; set; }

  public List<SurveyAnswer> Answers { get; set; }

  public string LastSyncDate { get; set; }

  public int UnopenedCount { get; set; }
}
```

The synchronization tasks run by the phone can be performed in both the background and the foreground. When synchronization occurs in the background, it is performed by a background agent. The background agent never runs its tasks while the Tailspin mobile client application is in the foreground. Therefore, the mobile client application and the background agent will never be concurrently attempting to update the survey data in isolated storage. For more information about background agents, see the section, "Synchronizing Data between the Phone and the Cloud," later in this chapter.

A background agent allows code to run in the background, even when the application is not running in the foreground. A background agent can be registered as a PeriodicTask, a ResourceIntensiveTask, or both.

The following code example shows how the **SurveyStore** class (that implements the **ISurveyStore** interface) performs the serialization and deserialization of the properties of the **SurveysList** instance to and from isolated storage.

```
C#
private readonly IsolatedStorageSettings isolatedStore;
private const string lastSyncDateKey = "lastSyncDateKey";
private const string unopenedCountKey = "unopenedCountKey";
private readonly string storeName;

public SurveyStore(string storeName)
{
  this.storeName = storeName;
  isolatedStore = IsolatedStorageSettings.ApplicationSettings;
  Initialize();
}

public SurveysList AllSurveys { get; set; }

...

private void SaveLastSyncDate()
{
  if (isolatedStore.Contains(lastSyncDateKey))
  {
    isolatedStore[lastSyncDateKey] = AllSurveys.LastSyncDate;
  }
  else
  {
    isolatedStore.Add(lastSyncDateKey, AllSurveys.LastSyncDate);
  }
  isolatedStore.Save();
}

...

private void SaveTemplates()
{
  using (var filesystem = IsolatedStorageFile.GetUserStoreForApplication())
  {
    using (var fs = new IsolatedStorageFileStream(
      storeName + ".templates", FileMode.Create, filesystem))
    {
      var serializer = new System.Runtime.Serialization
        .Json.DataContractJsonSerializer(typeof(List<SurveyTemplate>));
      serializer.WriteObject(fs, AllSurveys.Templates);
```

```
    }
  }
}

private void SaveAnswers()
{
  using (var filesystem = IsolatedStorageFile.GetUserStoreForApplication())
  {
    using (var fs = new IsolatedStorageFileStream(
      storeName + ".answers", FileMode.Create, filesystem))
    {
      var serializer = new System.Runtime.Serialization
        .Json.DataContractJsonSerializer(typeof(List<SurveyAnswer>));
      serializer.WriteObject(fs, AllSurveys.Answers);
    }
  }
}

public void SaveStore()
{
  SaveLastSyncDate();
  SaveTemplates();
  SaveAnswers();
}

...

private void Initialize()
{
  AllSurveys = new SurveysList();

  if (isolatedStore.Contains(lastSyncDateKey))
  {
    AllSurveys.LastSyncDate = (string)isolatedStore[lastSyncDateKey];
  }

  if (isolatedStore.Contains(unopenedCountKey))
  {
    AllSurveys.UnopenedCount = (int)isolatedStore[unopenedCountKey];
  }

  using (var filesystem = IsolatedStorageFile.GetUserStoreForApplication())
  {
    if (!filesystem.FileExists(storeName + ".templates"))
```

```
    {
      AllSurveys.Templates = new List<SurveyTemplate>();
    }
    else
    {
      using (var fs = new IsolatedStorageFileStream(
        storeName + ".templates", FileMode.Open, filesystem))
      {
        var serializer = new System.Runtime.Serialization
          .Json.DataContractJsonSerializer(typeof(
          List<SurveyTemplate>));
        AllSurveys.Templates = serializer.ReadObject(fs) as
          List<SurveyTemplate>;
      }
    }

    if (!filesystem.FileExists(storeName + ".answers"))
    {
      AllSurveys.Answers = new List<SurveyAnswer>();
    }
    else
    {
      using (var fs = new IsolatedStorageFileStream(
        storeName + ".answers", FileMode.Open, filesystem))
      {
        var serializer = new System.Runtime.Serialization
          .Json.DataContractJsonSerializer(typeof(List<SurveyAnswer>));
        AllSurveys.Answers = serializer.ReadObject(fs) as
          List<SurveyAnswer>;
      }
    }
  }
}
```

The following code example shows the **ISurveyStore** interface in the TailSpin.PhoneServices project that defines the operations that the application can perform on the survey store. The **SurveyStore** class implements this interface.

C#

```
public interface ISurveyStore
{
  string LastSyncDate { get; set; }
  SurveysList AllSurveys { get; set; }
  IEnumerable<SurveyTemplate> GetSurveyTemplates();
  IEnumerable<SurveyAnswer> GetCompleteSurveyAnswers();
```

```
  void SaveSurveyTemplates(IEnumerable<SurveyTemplate> surveys);
  void SaveSurveyAnswer(SurveyAnswer answer);
  SurveyAnswer GetCurrentAnswerForTemplate(
    SurveyTemplate template);
  void DeleteSurveyAnswers(
    IEnumerable<SurveyAnswer> surveyAnswers);
  void SaveStore();
  void SaveUnopenedCount();
  void MarkSurveyTemplateRead(SurveyTemplate template);

  event EventHandler SurveyAnswerSaved;
}
```

The **SurveyListViewModel** class calls the **GetSurveyTemplates** method to retrieve a list of surveys to display to the user. The **SurveyListViewModel** class creates filtered lists of surveys (new surveys, favorite surveys, and so on) after it has added the surveys to **CollectionViewSource** objects.

The **SurveysSynchronizationService** class uses the **GetCompleteSurveyAnswers**, **SaveSurveyTemplates**, and **DeleteSurveyAnswers** methods to manage the survey data stored on the device.

The **TakeSurveyViewModel** class uses the **GetCurrentAnswerForTemplate** and **SaveSurveyAnswer** methods to retrieve and save survey responses on the device. In addition, the **TakeSurveyViewModel** class uses the **GetSurveyTemplates** method to support pinned surveys.

Handling Activation and Deactivation

Windows Phone applications must be able to restore the UI state if the application is reactivated after the user has taken a call or used another application. In this scenario, the operating system makes the application dormant, tombstoned, and possibly terminates it, but it gives you the opportunity to save any data that you can use to put the application back in the same state if and when the operating system reactivates it.

Survey templates and survey answers are stored in separate files to enable future development of the synchronization service.

The application must restore its state if the user returns to the application after taking a call or using another application.

For more information about activation, deactivation, and tombstoning, see, *"Execution Model Overview for Windows Phone"* on MSDN.

The Windows Phone platform provides application-level activation and deactivation events, with the **Application_Activated** event handler in the **App** class notifying the application instance about whether it has resumed from dormancy or a tombstoned state.

The mobile client application is implemented using the MVVM pattern, with each view model being responsible for restoring its state when returning from a tombstoned state. When the mobile client application is reactivated, the information that indicates whether the application is returning from a tombstoned state is stored in the **ApplicationFrameNavigationService** class. The **ApplicationFrameNavigationService** class also provides functionality to determine if a page needs to recover from tombstoning, by using the **frame.BackStack** property. The base view model class then uses the **ApplicationFrameNavigationService** class to ensure that only pages that were tombstoned are resumed. In addition, the base view model class also provides two methods that can be overridden by child view models to implement logic that captures the UI state when deactivation occurs, and restores the UI state.

Overview of the Solution

The Windows Phone platform provides much of the infrastructure that you need to enable your application to restore the state of the UI when the application is reactivated:

- The phone uses a navigation service to facilitate navigating between pages. The **PhoneApplicationFrame** class exposes a **BackStack** property that can be used to build a dictionary which tracks the pages that were tombstoned.
- The phone uses a navigating event to notify the application when navigation is requested, and a navigated event to notify the application when the content that is being navigated to has been found and is available.
- The phone uses the application **Activated** event to notify the application when it has returned from deactivation. This event also informs the application about whether the instance was preserved, as is the case when returning from dormancy.
- The phone automatically records which screen your application is displaying, along with the current navigation stack, when it deactivates the application; then it rebuilds the navigation stack and redisplays this screen if the phone reactivates the application.

- The phone provides application-level and page-level state dictionaries in which you can store key/value pairs. These dictionaries are preserved when an application is tombstoned. When an application is activated after being tombstoned, these dictionaries are used to restore application and page state.

It's the application's responsibility to determine what state data it needs to save to be able to restore the application to the same state when the phone reactivates it.

In the Tailspin mobile client application, when the user navigates away from a page that gets recreated every time the page is visited, such as the AppSettingsView, no state is stored for that page. However, if the user navigates away from the page unintentionally, then the data required to recreate the page at a later time is stored in the application-level state dictionary, which survives tombstoning. For pages that do not get recreated every time they are visited, such as the SurveyListView, the data required to recreate the page is always stored in the application-level state dictionary irrespective of whether the navigation is intentional or unintentional. When the operating system reactivates the application, the operating system will redisplay the view that was active when the application was deactivated. When returning from dormancy, all object instances will still be in memory in the exact state prior to dormancy, so the application does not need to perform any additional tasks. However. when returning from a tombstoned state, the view model locator in the Tailspin mobile client application will instantiate the view model for the view, and the view model will restore its saved state and initialize any services so that it is back in the state it was in when it was originally deactivated.

The application should store only enough data to be able to restore the application to the same state it was in when it was deactivated. Also, remember it's possible that a deactivated application will never be reactivated, so you must also save any important data to permanent storage.

Inside the Implementation

Now is a good time to walk through the code that handles the activation, navigated and navigating events in more detail. As you go through this section, you may want to download the *Windows Phone Tailspin Surveys* application from the Microsoft Download Center.

The **Application_Activated** event handler in the **App** class notifies the application about whether it has resumed from dormancy or a tombstoned state. The following code example shows the **Application_Activated** event handler.

To meet the Windows Phone certification requirements, your application must complete the deactivation and reactivation processes within 10 seconds.

C#
```
private void Application_Activated(object sender,
  ActivatedEventArgs e)
{
  if (e.IsApplicationInstancePreserved)
  {
    PhoneApplicationService.Current.State.Clear();
  }
```

```
  else
  {
    ViewModelLocator.NavigationService.RecoveredFromTombstoning = true;
  }
}
```

The **IsApplicationInstancePreserved** property indicates whether the application instance was preserved intact in memory. If the **IsApplicationInstancePreserved** property of the **ActivatedEventArgs** is true, it means that the application has returned from dormancy and thus the application state dictionary is cleared. If the **IsApplicationInstancePreserved** property of the **ActivatedEventArgs** is false, it means that the application has returned from a tombstoned state and thus the **RecoveredFromTombstoning** property of the **ApplicationFrameNavigationService** instance is set to true.

The **ApplicationFrameNavigationService** class provides the **DoesPageNeedtoRecoverFromTombstoning** method, which determines if a page needs to recover from tombstoning. The following code example shows this method and the **UpdateTombstonedPageTracking** method.

```
C#
private Dictionary<string, bool> tombstonedPages;
...

public bool DoesPageNeedtoRecoverFromTombstoning(Uri pageUri)
{
  if (!RecoveredFromTombstoning) return false;

  if (tombstonedPages == null)
  {
    tombstonedPages = new Dictionary<string, bool>();
    tombstonedPages.Add(pageUri.ToString(), true);
    foreach (var journalEntry in frame.BackStack)
    {
      tombstonedPages.Add(journalEntry.Source.ToString(), true);
    }
    return true;
  }

  if (tombstonedPages.ContainsKey(pageUri.ToString()))
  {
    return tombstonedPages[pageUri.ToString()];
  }
  return false;
}

public void UpdateTombstonedPageTracking(Uri pageUri)
{
  tombstonedPages[pageUri.ToString()] = false;
}
```

The **DoesPageNeedtoRecoverFromTombstoning** method is used by the **ViewModel** class in order to determine whether or not to call the **OnPageResumeFromTombstoning** method in the **ViewModel** class. The method makes the assumption that the list of pages that were tombstoned can be determined by taking the first page that was instantiated due to a tombstoning recovery, and then examining the backstack. It uses the **BackStack** property of the **PhoneApplicationFrame** class to populate a dictionary with the pages that were tombstoned and uses this dictionary to track whether the **ViewModel** class called the **OnPageResumedFromTombstoning** method for each tombstoned page. The **UpdateTombstonedPageTracking** method simply updates a given page entry in the tombstonedPages dictionary to mark that the page has completed tombstone recovery.

A facade is a design pattern that provides a simplified interface for a class. The facade translates calls to its interface into calls to the original class. This approach enables writing loosely coupled code that is testable.

The developers at Tailspin created the **IPhoneApplicationServiceFacade** interface which is implemented by the **PhoneApplicationServiceFacade** class. This facade provides a simplified interface to the **PhoneApplicationService** class. The following code example shows the **IPhoneApplicationServiceFacade** interface.

A service is a facade that exposes a loosely coupled unit of functionality that implements one action.

C#
```
public interface IPhoneApplicationServiceFacade
{
  void Save(string key, object value);
  T Load<T>(string key);
  void Remove(string key);
}
```

The **PhoneApplicationServiceFacade** class has methods that save and load any state that the page needs when it's navigated to or navigated away from during application deactivation. The following code example shows the complete **PhoneApplicationServiceFacade** class.

There is also a **PhoneApplicationPage** class that allows you to store transient page state as you navigate away from a page and restore the state when you return to the page.

C#
```
public class PhoneApplicationServiceFacade :
  IPhoneApplicationServiceFacade
{
  public void Save(string key, object value)
  {
    if (PhoneApplicationService.Current.State.ContainsKey(key))
    {
      PhoneApplicationService.Current.State.Remove(key);
    }
```

```
      PhoneApplicationService.Current.State.Add(key, value);
    }

    public T Load<T>(string key)
    {
      object result;

      if (!PhoneApplicationService.Current.State
        .TryGetValue(key, out result))
      {
        result = default(T);
      }
      else
      {
        PhoneApplicationService.Current.State.Remove(key);
      }

      return (T)result;
    }

    public void Remove(string key)
    {
      if (PhoneApplicationService.Current.State.ContainsKey(key))
      {
        PhoneApplicationService.Current.State.Remove(key);
      }
    }
}
```

Each view model is responsible for managing its own state when the page is navigated to or navigated away from, either intentionally or during application deactivation. Most of the view models in the application derive from the **ViewModel** class that listens to the **Navigated** and **Navigating** events in the **ApplicationFrameNavigationService** class, which is a facade over the **PhoneApplicationFrame** SDK class. The base **ViewModel** class also uses the **ApplicationFrameNavigationService** class to determine if the application is returning from a tombstoned state. The following code example shows part of the abstract **ViewModel** class.

The objects that you save in the State dictionary must be serializable.

The developers at Tailspin chose to make each view model responsible for persisting and reloading its own state.

C#

```
public abstract class ViewModel : NotificationObject, IDisposable
{
  private readonly INavigationService navigationService;
  private readonly IPhoneApplicationServiceFacade
    phoneApplicationServiceFacade;
  private bool disposed;
  private readonly Uri pageUri;
  private static Uri currentPageUri;

  protected ViewModel(INavigationService navigationService,
    IPhoneApplicationServiceFacade phoneApplicationServiceFacade,
    Uri pageUri)
  {
    this.pageUri = pageUri;
    this.navigationService = navigationService;
    this.phoneApplicationServiceFacade = phoneApplicationServiceFacade;

    this.navigationService.Navigated +=
      this.OnNavigationService_Navigated;
    this.navigationService.Navigating +=
      this.OnNavigationService_Navigating;
  }

  void OnNavigationService_Navigating(object sender,
    System.Windows.Navigation.NavigatingCancelEventArgs e)
  {
    if (currentPageUri == null || pageUri == null) return;

    if (currentPageUri.ToString().StartsWith(pageUri.ToString()))
    {
      OnPageDeactivation(e.IsNavigationInitiator);
    }
  }

  void OnNavigationService_Navigated(object sender,
    System.Windows.Navigation.NavigationEventArgs e)
  {
    if (IsResumingFromTombstoning)
    {
        if (e.Uri.ToString().StartsWith(pageUri.ToString()))
        {
            OnPageResumeFromTombstoning();
            navigationService.UpdateTombstonedPageTracking(pageUri);
        }
```

```
    }
    currentPageUri = e.Uri;
  }

  ...

  protected bool IsResumingFromTombstoning
  {
    get
    {
      return navigationService.DoesPageNeedToRecoverFromTombstoning(pageUri);
    }
  }

  ...

  public IPhoneApplicationServiceFacade
    PhoneApplicationServiceFacade
  {
    get { return this.phoneApplicationServiceFacade; }
  }

  public virtual void OnPageDeactivation(bool isIntentionalNavigation)
  {
  }

  public abstract void OnPageResumeFromTombstoning();

  ...

  protected virtual void Dispose(bool disposing)
  {
    ...
    if (disposing)
    {
       navigationService.Navigated -=
         this.OnNavigationService_Navigated;
       navigationService.Navigating -=
         this.OnNavigationService_Navigating;
    }
    ...
  }
}
```

The **OnNavigationService_Navigating** event handler calls the **OnPageDeactivation** method only on the view model of the page where navigation is requested. The **IsNavigationInitiator** property of the **NavigatingCancelEventArgs** class is passed as a parameter into the **OnPageDeactivation** method. This property indicates whether the current application is the both the origin and destination of the navigation. If it is, it means that page-based navigation within the application is occurring. In this case, deactivation will not be performed. If the current application is not the origin and destination of the navigation (for instance the Windows Phone device has received a phone call), deactivation will occur and the state of the page will be stored in the application state. However, the **SelectedPivotIndex** property of the **SurveyListViewModel** class is stored even when the user intentionally navigates away from the page. This is because the mobile client application may be deactivated from another page, such as the FilterSettingsView page, and when the application is reactivated, that page will be restored. However, when the user navigates back to the SurveyListView page, the **SelectedPivotIndex** property must be restored so that the user sees the survey list that was being viewed before the application was deactivated.

Remember, the navigating event notifies the application when navigation is requested, and the navigated event notifies the application when the content that is being navigated to has been found and is available.

The **OnNavigationService_Navigated** event handler is used to determine if the view model instance needs to recover from tombstoning. It first checks the **IsResumingFromTombstoning** property to determine whether the application is returning from a tombstoned state, and whether the view model has already recovered its tombstoned state. If the **IsResumingFromTombstoning** property is true, the application is returning from tombstoning. If the **IsResumingFromTombstoning** property is false, the application is either returning from dormancy or undertaking a page-based navigation. If the application is returning from tombstoning, the **OnPageResumeFromTombstoning** method is called on the view model of the page that's been navigated to, to restore the desired state to the view model. The **ApplicationFrameNavigationService** is then notified that the page has completed tombstone recovery, thus ensuring that pages that were tombstoned are resumed only once.

Each view model can override the **OnPageResumeFromTombstoning** and **OnPageDeactivation** methods to provide its custom state saving and restoring behavior. The following code example shows how the **SurveyListViewModel** class saves and restores its state.

```
C#
public SurveyListViewModel(
  ISurveyStoreLocator surveyStoreLocator,
  ISurveysSynchronizationService synchronizationService,
  INavigationService navigationService,
  IPhoneApplicationServiceFacade phoneApplicationServiceFacade,
  IShellTile shellTile,
  ISettingsStore settingsStore,
  ILocationService locationService)
  : base(navigationService, phoneApplicationServiceFacade,
        new Uri(@"/Views/SurveyList/SurveyListView.xaml", UriKind.Relative))
{
  ...
}

...

public override void OnPageDeactivation(bool isIntentionalNavigation)
{
  this.PhoneApplicationServiceFacade.Save("MainPivot",
    this.SelectedPivotIndex);
}

public override sealed void OnPageResumeFromTombstoning()
{
  this.selectedPivotIndex =
    this.PhoneApplicationServiceFacade.Load<int>("MainPivot");
}
```

The **OnPageDeactivation** method is called by the **Navigating** event of the **PhoneApplication-Frame** class when the page is intentionally or unintentionally navigated away from. It uses the **Save** method from the **PhoneApplicationServiceFacade** class to save the **MainPivot** key/value pair to application state, with the value being the **SelectedPivotIndex** property value.

The **OnPageResumeFromTombstoning** method is called by the **Navigated** event of the **Phone-ApplicationFrame** class when the page is resuming from tombstoning. It uses the **Load** method from the **PhoneApplicationServiceFacade** class to load the MainPivot key/value pair from the application state and store the value in the **selectedPivotIndex** field.

Reactivation and the Pivot Control

When your application is reactivated by the operating system, you should restore the UI state, which in the Tailspin application includes displaying the active question if the application was made dormant or tombstoned while the user was completing a survey. The following code example shows the **OnPageResumeFromTombstoning** method of the **TakeSurveyViewModel** class, which restores the **SelectedPivotIndex** property, along with the survey answer and survey id, from the application state, when the application returns from a tombstoned state.

If the application has been deactivated, and the user relaunches it from Start, the state data is discarded.

C#
```
public override sealed void OnPageResumeFromTombstoning()
{
  this.tombstoned = this.PhoneApplicationServiceFacade
    .Load<SurveyAnswer>("TakeSurveyAnswer");
  this.SelectedPivotIndex = this.PhoneApplicationServiceFacade
    .Load<int>("TakeSurveyCurrentIndex");
  this.surveyId = this.PhoneApplicationServiceFacade
    .Load<string>("TakeSurveyId");
  Initialize(this.surveyId);
  this.locationService.StartWatcher();
}
```

This method is called by the **Navigated** event of the **PhoneApplicationFrame** class when the page is resuming from tombstoning. It uses the **Load** method from the **PhoneApplicationServiceFacade** class to load from application state the **TakeSurveyAnswer** key/value pair, the **TakeSurveyCurrentIndex** key/value pair, and the **TakeSurveyId** key/value pair. It then calls the **Initialize** method, passing the surveyId field into the method.

To display the correct question when the application is reactivated, Tailspin changes the **SelectedItem** property of the control. The following code example shows the event handlers in the code-behind for the **TakeSurveyView** page.

C#
```
private bool loaded;

private void PivotSelectionChanged(object sender,
  System.Windows.Controls.SelectionChangedEventArgs e)
{
  if (this.loaded)
  {
    ((TakeSurveyViewModel)this.DataContext)
      .SelectedPivotIndex = this.questionsPivot.SelectedIndex;
  }
```

```
}

private void ControlLoaded(object sender,
  System.Windows.RoutedEventArgs e)
{
  var vm = (TakeSurveyViewModel)this.DataContext;
  this.questionsPivot.SelectedItem =
    this.questionsPivot.Items[vm.SelectedPivotIndex];
  this.loaded = true;
  ...
}
```

The **SelectedPivotIndex** of the **TakeSurveyViewModel** class tracks the currently active question in the **Pivot** control.

Handling Asynchronous Interactions

Chapter 2, "Building the Mobile Client," describes how Tailspin implemented commands in the mobile client application. For some commands, Tailspin implements the command asynchronously to avoid locking the UI while a time-consuming operation is running.

For example, on the AppSettingsView page, a user can enable or disable push notifications of new surveys from the Microsoft Push Notification Service (MPNS). This requires the application to send a request to the MPNS that the application must handle asynchronously. The application displays a progress indicator on the AppSettingsView page while it handles the asynchronous request.

For more information about MPNS, see Chapter 4, "Connecting with Services."

Tailspin decided to use the Reactive Extensions (Rx) for .NET to run asynchronous tasks on the phone because it enables them to create compact, easy-to-understand code for complex asynchronous operations.

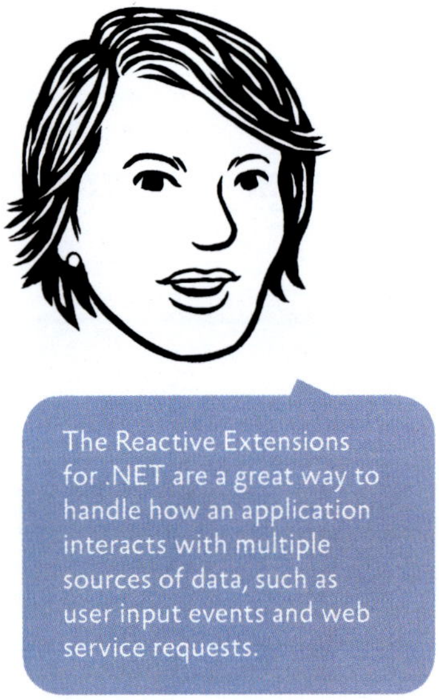

Using Reactive Extensions

Rx allows you to write compact, declarative code to manage complex, asynchronous operations. Rx can be understood by comparing it to the more familiar concept of enumerable collections. Figure 3 shows two alternative approaches to iterating over a sequence.

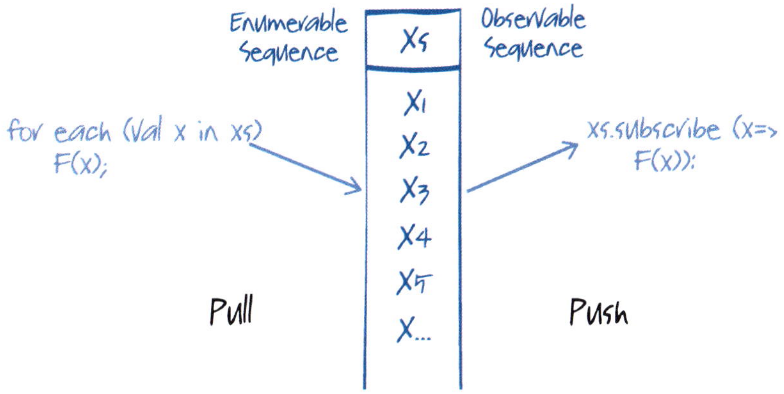

Figure 3
Enumerable and observable sequences

To iterate over an enumerable sequence, you can use an iterator to pull each item from the sequence in turn, which is what the C# **foreach** construct does for you. With an observable sequence, you subscribe to an observable object that pushes items from the sequence to you. For example, you can treat the events raised by a control or the data arriving over the network as an observable sequence. Furthermore, you can use standard LINQ operators to filter the items from the observable sequence, and control which thread you use to process each item as it arrives.

Inside the Implementation

The application performs an asynchronous request to the Microsoft Push Notification Service when the user subscribes to push notifications on the AppSettingsView page. The following code example from the AppSettingsView.xaml file shows the definitions of the progress indicator that is active during the asynchronous request and the **ToggleSwitch** control that enables the user to set his or her preference.

XAML

```
...
xmlns:toolkit="clr-namespace:Microsoft.Phone.Controls;
assembly=Microsoft.Phone.Controls.Toolkit"
...
<shell:SystemTray.ProgressIndicator>
  <shell:ProgressIndicator IsIndeterminate="True"
    IsVisible="{Binding IsSyncing}"
    Text="{Binding ProgressText}"/>
</shell:SystemTray.ProgressIndicator>
...

<toolkit:ToggleSwitch Header="Subscribe to Push Notifications"
  Margin="0,202,0,0"
  IsChecked="{Binding SubscribeToPushNotifications, Mode=TwoWay}" />
```

The **ToggleSwitch** control binds to the **SubscribeToPushNotifications** property of the **AppSettingsViewModel**, and the **ProgressIndicator** control binds to the **IsSyncing** property of the **AppSettingsViewModel** class.

The following code example from the **AppSettingsViewModel** class shows what happens when the user clicks the **Save** button on the AppSettingsView page. This **Submit** method handles the asynchronous call to the **UpdateReceiveNotifications** method in the **RegistrationServiceClient** class by using Rx. Before it calls the **UpdateReceiveNotifications** method, it first sets the **IsSyncing** property to **true** so that the UI can display an active progress meter. The method handles the asynchronous call in three steps:

1. The **UpdateReceiveNotifications** method in the **RegistrationServiceClient** class returns an observable **TaskSummaryResult** object that contains information about the task.
2. The **Submit** method uses the **ObserveOnDispatcher** method to handle the **TaskSummaryResult** object on the dispatcher thread.
3. The **Subscribe** method specifies how to handle the **TaskSummaryResult** object and starts the execution of the source observable sequence by asking for the next item.

```
C#
private readonly IRegistrationServiceClient registrationServiceClient;
private IDisposable subscription;
...

public void Submit()
{
  ...
  this.isSyncing = true;
  ...

  if (this.SubscribeToPushNotifications ==
    this.settingsStore.SubscribeToPushNotifications)
  {
    this.IsSyncing = false;
    if (this.NavigationService.CanGoBack) this.NavigationService.GoBack();
    return;
  }

  ...

  subscription = this.registrationServiceClient.
    .UpdateReceiveNotifications(this.SubscribeToPushNotifications)
    .ObserveOnDisptacher()
    .Subscribe
    (
        taskSummary =>
        ... ,
```

```
        exception =>
        ...
    );
}
```

The following code example shows the definition of the action that the **Subscribe** method performs when it receives a **TaskSummaryResult** object. If the **TaskSummaryResult** object indicates that the change was successful, it updates the setting in local isolated storage, sets the **IsSyncing** property to false, and navigates back to the previous view. If the **TaskSummaryResult** object indicates that the change failed, it reports the error to the user.

C#

```
taskSummary =>
{
  if (taskSummary == TaskSummaryResult.Success)
  {
    this.settingsStore.SubscribeToPushNotifications =
      this.SubscribeToPushNotifications;
    this.IsSyncing = false;
    if (this.NavigationService.CanGoBack) this.NavigationService.GoBack();
    if (!SubscribeToPushNotifications)
    {
      CleanUp();
    }
  }
  else
  {
    // Update unsuccessful, probably due to communication issue with
    // Registration Service. Don't close channel so that we can retry later.
    if (!SubscribeToPushNotifications)
    {
      CleanUp();
    }
    var errorText = TaskCompletedSummaryStrings
      .GetDescriptionForResult(taskSummary);
    this.IsSyncing = false;
    this.submitErrorInteractionRequest.Raise(
      new Notification
      {
        Title = "Push Notification: Server error",
        Content = errorText
      },
      n => { });
  }
  this.CanSubmit = true;
}
```

The **Subscribe** method can also handle an exception returned from the asynchronous task. The following code example shows how it handles the scenario in the Tailspin mobile client where the asynchronous action throws a **WebException** exception.

C#
```
exception =>
{
  this.CanSubmit = true;

  // Update unsuccessful, probably due to communication issue
  // with Registration Service. Don't close channel so that
  // we can retry later.
  if (SubscribeToPushNotifications)
  {
    CleanUp();
  }

  if (exception is WebException)
  {
    var webException = exception as WebException;
    var summary = ExceptionHandling.GetSummaryFromWebException(
      "Update notifications", webException);
    var errorText = TaskCompletedSummaryStrings
      .GetDescriptionForResult(summary.Result);
    this.IsSyncing = false;
    this.submitErrorInteractionRequest.Raise(
     new Notification
     {
       Title = "Push Notification: Server error",
       Content = errorText
     },
     n => { });
  }
  else
  {
    throw exception;
  }
}
```

Synchronizing Data between the Phone and the Cloud

The Tailspin mobile client synchronizes survey definitions and answers between the phone and the Tailspin cloud service.

The Tailspin mobile client must be able to download new surveys from the Tailspin Surveys service and upload survey answers to the service. This section describes how Tailspin designed and implemented this functionality. It focuses on the details of the synchronization logic instead of on the technologies the application uses to store data locally and to interact with the cloud. Details of the local storage implementation are described earlier in this chapter, and Chapter 4, "Connecting with Services," describes how the mobile client application interacts with Tailspin's cloud services.

There are two separate synchronization tasks that the mobile client must perform:

- The mobile client must download from the cloud service any new surveys that match the user's subscription criteria.
- The mobile client must send completed survey answers to the cloud service for analysis.

These two tasks are independent of each other; therefore, the mobile client can perform these operations in parallel. Furthermore, for the Tailspin application, the synchronization logic is very simple. At the time of this writing, the Tailspin cloud application does not allow subscribers to modify or delete their survey definitions, so the mobile client only needs to look for new survey definitions. On the client, a surveyor cannot modify survey answers after the survey is complete, so the mobile client can send all of its completed survey answers to the cloud service and then remove them from the mobile client's local store.

Tailspin's synchronization logic is relatively simple. A more complex client application may have to deal with modified and deleted data during the synchronization process.

In the Tailspin mobile client application, the synchronization process can be initiated automatically or manually by the user tapping a button. Because synchronization can be a time-consuming process, the mobile client should perform synchronization asynchronously, and notify the user of the outcome when the synchronization completes.

How often you should run a synchronization process in your application involves some trade-offs. More frequent synchronizations mean that the data on both the client and in the service is more up to date. It can also help to free up valuable storage space on the client if the client no longer needs a local copy of the data after it has been transferred to the service. Data stored in the service is also less vulnerable to loss or unauthorized access. However, synchronization is often a resource-intensive process itself, consuming battery power and CPU cycles on the mobile client and using potentially expensive bandwidth to transfer the data. You should design your synchronization logic to transfer as little data as possible.

Tailspin offers both manual and automatic synchronization between the phone and the cloud.

Overview of the Solution

Tailspin considered using the Microsoft Sync Framework, but they decided to implement the synchronization logic themselves. The reason for this decision was that the synchronization requirements for the application are relatively simple, which meant that the risks associated with developing this functionality themselves was lower. The developers at Tailspin have designed the synchronization service so that they can easily replace the synchronization functionality with an alternative implementation in the future.

Automatic Synchronization

Automatic synchronization between the mobile client application and the Surveys cloud application is performed by a background agent. Background agents allow an application to execute code in the background, even when the application is not running in the foreground. Background agents can run two types of task:

- Periodic tasks that run for a short period of time at regular intervals. A typical scenario for this type of task is performing small amounts of data synchronization.
- Resource-intensive tasks that run for a relatively long period of time when the phone meets a set of requirements relating to processor activity, power source, and network connection. A typical scenario for this type of task is synchronizing large amounts of data to the phone while it is not actively being used.

An application may have only one background agent, which must be registered as a periodic task, a resource-intensive task, or both. The schedule on which the agent runs depends on which type of task is registered.

Periodic tasks typically run for up to 25 seconds every 30 minutes. Other constraints may prevent a periodic task from running.

Resource-intensive tasks typically run for up to 10 minutes. In order to run, the Windows Phone device must be connected to an external power source and have a battery power greater than 90%. In addition, the Windows Phone device must have a network connection over Wi-Fi or through a connection to a PC, and the device screen must be locked.

The mobile client application uses a periodic task to download any new surveys that match the user's subscription criteria, and a resource-intensive task to upload completed survey answers to the cloud service. The upload only occurs if certain constraints are met on the device. A toast notification is used to inform the user of the result of a background task when it is performed.

The scenarios that control the lifespan of the background tasks are as follows:

- When the application launches, the periodic task and the resource-intensive task are removed from the operating system scheduler.
- When the application closes, the periodic task and the resource-intensive task are added to the operating system scheduler.

This design decision ensures that both the periodic task and the resource-intensive task will never run synchronization tasks in the background while the application is running synchronization tasks is in the foreground, thus avoiding any potential concurrency issues.

The background tasks use Rx to perform the synchronization. However, there is no guarantee that the tasks will ever run, due to restrictions such as battery life, network connectivity, and memory use. Therefore, it is still possible for the user to initiate synchronization manually. For more information about background agents, see, "Background Agents Overview for Windows Phone."

Manual Synchronization

Tailspin decided to use the Rx to run the two manual synchronization tasks asynchronously and in parallel on the phone. Figure 4 summarizes the manual synchronization process and the tasks that it performs.

It is possible that a resource-intensive agent will never be run on a particular phone, due to the phone constraints that must be met. You should consider this when designing your application. It may be more appropriate to use the background file transfer service if the data to be transferred can be grouped into separate files that can be queued up.

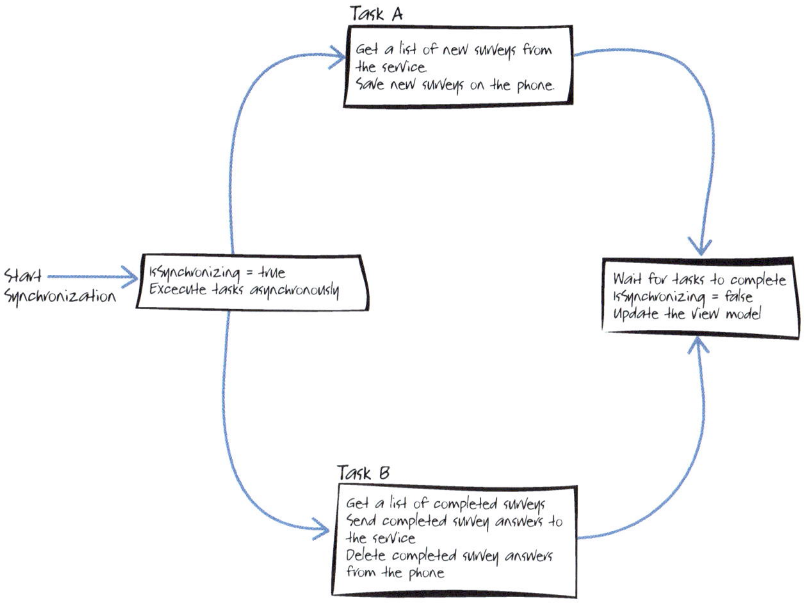

FIGURE 4
The manual synchronization process on the phone

The user starts the synchronization process by tapping a button in the UI. A progress indicator in the UI is bound to the IsSynchronizing property in the view model to provide a visual cue that the synchronization process is being performed. Rx runs the two tasks in parallel, and after both tasks complete, it updates the view model with the new survey data.

In Figure 4, Task A is responsible for downloading a list of new surveys for the user and saving them locally in isolated storage. The service creates the list of new surveys to download based on information sent by the mobile client application. The client sends the date of the last synchronization so that the service can find surveys created since that date, and the service uses the user name sent by the client to filter for surveys that the user is interested in. For more information, see the section, "Filtering Data," in Chapter 4, "Connecting with Services."

It's important to let the user know that an operation is running asynchronously. When you don't know how long it will take, use the indeterminate progress bar.

Task B sends all completed survey answer data to the cloud service, and then it deletes the local copy to free up storage space on the phone.

When both tasks are complete, the application updates the data in the view model, the UI updates based on the bindings between the view and the view model, and the application displays a toast notification if the synchronization was successful or an error pop-up window otherwise. For more information about how the mobile client application handles UI notifications, see the section, "User Interface Notifications," in Chapter 2, "Building the Mobile Client."

Limitations of the Current Approach

As discussed earlier, Tailspin's requirements for the synchronization service are relatively simple because the online Tailspin Surveys service does not allow tenants to modify a survey after they have published it. However, it is possible for tenants to delete surveys in the online application. The current synchronization process in the sample application does not take this into account, so the number of survey definitions stored on the client never decreases. Furthermore, the client will continue to be able to submit answers to surveys that no longer exist in the online service. A real implementation should extend the synchronization logic to accommodate this scenario. One possible solution would be to give every survey an expiration date and make it the mobile client's responsibility to remove out-of-date surveys. Another solution would be to adopt a full-blown synchronization service, such as the Microsoft Sync Framework.

In addition, the current approach does not address the use case where a user removes a tenant from their list of preferred tenants. The mobile client application will not receive any new surveys from the deselected tenants, but the application does not remove any previously downloaded surveys from tenants who are no longer on the list. A complete synchronization solution for Tailspin should also address this use case.

Inside the Implementation

Now is a good time to walk through the code that implements the data synchronization in more detail. As you go through this section, you may want to download the *Windows Phone Tailspin Surveys* application from the Microsoft Download Center.

Automatic Synchronization

The **App** class controls the lifespan of the background tasks via the **Application_Launching** and **Application_Closing** methods. The following code example shows these methods.

```
C#
// Code to execute when the application is launching
// (e.g., from Start)
// This code will not execute when the application is reactivated
private void Application_Launching(object sender,
  LaunchingEventArgs e)
{
  this.ViewModelLocator.ScheduledActionClient.ClearTasks();
}

// Code to execute when the application is closing
// (e.g., user hit Back)
// This code will not execute when the application is deactivated
private void Application_Closing(object sender,
  ClosingEventArgs e)
{
  this.ViewModelLocator.SurveyListViewModel
    .ResetUnopenedSurveyCount();

  this.ViewModelLocator.ScheduledActionClient.EnsureTasks();
  this.ViewModelLocator.Dispose();
}
```

The **Application_Launching** method calls the **ClearTasks** method from the **ScheduledActionClient** class. This will remove both the periodic task and the resource-intensive task from the operating system scheduler. The **Application_Closing** method calls the **ResetUnopened-SurveyCount** method of the **SurveyListViewModel** class, which resets the **UnopenedSurveyCount** property of the **SurveysSynchronization-Service** class and the **Count** property of the Application Tile. It then calls the **EnsureTasks** method from the **ScheduledActionClient** class, which adds both the periodic task and the resource-intensive task to the operating system scheduler before disposing of the instance of the **ViewModelLocator** class. However, the **Application_Closing** method is only executed when the user navigates backwards past the first page of the application. Therefore, if the user does not exit the application by using this approach, the background tasks will not be added to the operating system scheduler.

A resource-intensive task is used to upload completed surveys to the cloud service, as the surveys answers could include audio and images.

The same background tasks run for all users. Changing the username does not create new background tasks.

The **ScheduledActionClient** class implements the **IScheduledActionClient** interface and provides a facade over the **ScheduledActionServiceAdapter** class, which in turn adapts the **ScheduledActionService** class from the API. The purpose of adapting the **ScheduledActionService** class is to create a loosely coupled class that is testable. The following code example shows the **UserEnabled** property from the **ScheduledActionClient** class, along with the **AddPeriodicTask**, **AddResourceIntensiveTask**, **ClearTasks**, and **EnsureTasks** methods.

C#

```
private readonly IScheduledActionService scheduledActionService;

...

public void AddPeriodicTask(string taskName, string taskDescription,
  TimeSpan debugDelay)
{
  RemoveTask(taskName);

  var periodicTask = new PeriodicTask(taskName);
  periodicTask.Description = taskDescription;

  scheduledActionService.Add(periodicTask);
#if DEBUG
  if (debugDelay > TimeSpan.Zero)
    scheduledActionService.LaunchForTest(taskName, debugDelay);
#endif
}

public void AddResourceIntensiveTask(string taskName, string taskDescription,
  TimeSpan debugDelay)
{
  RemoveTask(taskName);

  var resourceIntensiveTask = new ResourceIntensiveTask(taskName);
  resourceIntensiveTask.Description = taskDescription;

  scheduledActionService.Add(resourceIntensiveTask);

#if DEBUG
  if (debugDelay > TimeSpan.Zero)
    scheduledActionService.LaunchForTest(taskName, debugDelay);
#endif
}

public void ClearTasks()
{
  RemoveTask(Constants.PeriodicTaskName);
```

```
    //removed only because this sample will normally be reviewed in a debug scenario
    //where the resource-intensive task may run while the app is in the foreground
    //possibly creating concurrency issues
    RemoveTask(Constants.ResourceTaskName);
  }

  public void EnsureTasks()
  {
    if (UserEnabled())
    {
      try
      {
        AddPeriodicTask(Constants.PeriodicTaskName,
          Constants.PeriodicTaskDescription, TimeSpan.FromMinutes(3));
        AddResourceIntensiveTask(Constants.ResourceTaskName,
          Constants.ResourceTaskDescription, TimeSpan>FromMinutes(3));
      }
      catch
      {
        //possible exception is hidden here since this method is called
        //during app closing. Check for OS-level disabling of background tasks
        //is checked when saving on the Settings page
      }
    }
  }

  public bool IsEnabled
  {
    get
    {
      bool result = true;

      try
      {
        //currently the only way to check if a user has disabled background agents
        //at the OS settings level is to attempt to add them
        AddPeriodicTask(Constants.PeriodicTaskName,
         Constants.PeriodicTaskDescription);
        RemoveTask(Constants.PeriodicTaskName);
      }
      catch (InvalidOperationException exception)
      {
        if (exception.Message.Contains(Constants.DisabledBackgroundException))
        {
```

```
        result = false;
      }
    }

    return result;
  }
}

public bool UserEnabled
{
  get
  {
    return !string.IsNullOrEmpty(settingsStore.UserName) &&
      settingsStore.BackgroundTasksAllowed;
  }
}
...
```

The **UserEnabled** property returns a Boolean value indicating whether or not the settings store contains a username, and whether or not background tasks are turned on in the application. The **AddPeriodicTask** method adds a new periodic task to the operating system scheduler by calling the **Add** method of the **ScheduledActionServiceAdapter** class, which in turn calls the **ScheduledActionService Add** method from the API. Similarly, a new resource-intensive task is added to the operating system scheduler by the **AddResourceIntensiveTask** method. The **ClearTasks** method is called when the application launches, and removes both the periodic task and the resource-intensive task from the operating system scheduler. The **EnsureTasks** method is called when the application closes, and adds both the periodic task and the resource-intensive task to the operating system scheduler. The **IsEnabled** property checks if the user has disabled background agents in the operating system settings. It does this by attempting to add a **PeriodicTask**, and then removes it. If an **InvalidOperationException** occurs, it means that background agents are disabled in the operating system settings.

The methods that execute the background tasks are contained in the **ScheduledAgent** class in the TailSpin.PhoneAgent project. The following code example shows the **OnInvoke** method, which executes the background tasks.

C#

```
protected override void OnInvoke(ScheduledTask task)
{
  if (task is PeriodicTask)
  {
    RunPeriodicTask(task);
  }
  else if(task is ResourceIntensiveTask)
  {
    RunResourceIntensiveTask(task);
  }
}
```

The **OnInvoke** method accepts a **ScheduledTask** as a parameter, and if it's a **PeriodicTask**, calls the **RunPeriodicTask** method. If the parameter is a **ResourceIntensiveTask**, the **RunResourceIntensive-Task** method is called. The following code example shows the **RunPeriodicTask** method.

```
C#
private void RunPeriodicTask(ScheduledTask task)
{
#if ONLY_PHONE
  var surveyServiceClient = new SurveysServiceClientMock(settingsStore);
#else
  var httpClient = new HttpClient();
  var surveyServiceClient = new SurveysServiceClient(
    new Uri("http://127.0.0.1:8080/Survey/"), settingsStore, httpClient);
#endif
  var surveyStoreLocator = new SurveyStoreLocator(settingsStore,
    storeName => new SurveyStore(storeName));
  var synchronizationService = new SurveysSynchronizationService(
    () => surveyServiceClient, surveyStoreLocator);

  synchronizationService
    .GetNewSurveys()
    .ObserveOnDispatcher()
    .Subscribe(SyncCompleted, SyncFailed);

#if DEBUG
  ScheduledActionService.LaunchForTest(task.Name, TimeSpan.FromMinutes(3));
#endif
}
```

The **RunPeriodicTask** method uses Rx to run the synchronization process asynchronously. The asynchronous calls are handled as follows:

1. The **GetNewSurveys** method in the **SurveysSynchronizationService** class returns an observable **TaskSummaryResult** object that contains information about the task.
2. The **RunPeriodicTask** method uses the **ObserveOnDispatcher** method to handle the **TaskSummaryResult** object on the dispatcher thread.
3. The **Subscribe** method specifies how to handle the **TaskSummaryResult** object and how to handle an error occurring.

The **LaunchForTest** method is used to launch the background agent when debugging. Periodic agents are not launched by the system when the debugger is attached. This method can be called from the foreground application while debugging, enabling you to step through the background agent code.

Memory and execution time policies are not enforced while the debugger is attached. Therefore, it is important that you test your agent while not debugging to verify that your agent does not exceed the memory cap or run longer than the allotted time period for the agent type.

The following code example shows the definition of the action that the Subscribe method performs when it receives a **TaskSummaryResult** object.

```
C#
private void SyncCompleted(TaskCompletedSummary taskSummary)
{
  int newCount;

  if (taskSummary != null &&
      int.TryParse(taskSummary.Context, out newCount) &&
      newCount > 0)
  {
    var toast = new ShellToast();
    toast.Title = TaskCompletedSummaryStrings
     .GetDescriptionForSummary(taskSummary);
    toast.Content = "";
    toast.Show();
  }

  NotifyComplete();
}
```

If the **TaskSummaryResult** object indicates that new surveys have been downloaded, a toast notification is built that informs the user that synchronization was successful and indicates how many new surveys have been downloaded. Alternatively, if the **TaskSummaryResult** object indicates that completed survey answers have been uploaded, a toast notification is built that informs the user that synchronization was successful and indicates how many survey's answers were uploaded. The **BackgroundAgent.NotifyComplete** method is then called to inform the operating system that the agent has completed its intended task for the current invocation of the agent.

The **Subscribe** method can also handle an exception returned from the asynchronous task. The following code example shows how it handles the scenario where the asynchronous action throws an exception.

```
C#
private void SyncFailed(Exception ex)
{
  Abort();
}
```

The **SyncFailed** method simply calls the **BackgroundAgent.Abort** method to inform the OS that the agent is unable to perform its intended task and that it should not be launched again until the foreground application mitigates the blocking issues and re-enables the agent.

The **RunResourceIntensiveTask** method uses a similar approach to the one outlined here for the **RunPeriodicTask** method.

Manual Synchronization

The user can also initiate the synchronization process by tapping the Sync button on the SurveyListView page. This sends a command to the SurveyListViewModel view model which, in turn, starts the synchronization process. While the synchronization process is running, the application displays an indeterminate progress indicator because it has no way of telling how long the synchronization process will take to complete. If the synchronization process is successful, the SurveyListViewModel class rebuilds the lists of surveys that are displayed by the SurveyListView page. If the synchronization process fails with a network error or a credentials error, the SurveyListViewModel class does not rebuild the lists of surveys that are displayed by the SurveyListView page.

For information about how the user initiates the synchronization process from the user interface, see the section "Commands" in Chapter 2, "Building the Mobile Client."

The **SurveyListViewModel** class uses Rx to run the synchronization process asynchronously by invoking the **StartSynchronization** method in the **SurveysSynchronizationService** class. When the synchronization process is complete, the **SurveysSynchronizationService** class returns a summary of the synchronization task as a collection of **TaskCompletedSummary** objects. The view model updates the UI by using the **Observe-OnDispatcher** method to run the code on the dispatcher thread. The following code example shows the **StartSync** method in the **SurveyListViewModel** class that interacts with the **SurveysSynchronizationService** class.

```
C#
private readonly
  ISurveysSynchronizationService synchronizationService;
...
public void StartSync()
{
  if (this.IsSynchronizing)
  {
    return;
  }

  this.IsSynchronizing = true;
  this.synchronizationService
      .StartSynchronization()
      .ObserveOnDispatcher()
      .Subscribe(this.SyncCompleted);
}
```

Using Rx can make code that handles asynchronous operations simpler to understand and more compact.

The **SurveysSynchronizationService** class uses Rx to handle the parallel, asynchronous behavior in the synchronization process. Figure 5 shows the overall structure of the StartSync and **StartSynchronization** methods and how they use Rx to run the synchronization tasks in parallel.

The **StartSync** method (**SurveyListViewModel** class) subscribes to:

> The **StartSynchronization** method (**SurveysSynchronizationService** class) that declares:
>
> > The **getNewSurveys** task to get a list of new surveys from the Tailspin Surveys service and return: **IObservable<TaskCompletedSummary>.**
>
> > The **saveSurveyAnswers** task to save completed surveys to the Tailspin Surveys service and return: **IObservable<TaskCompletedSummary>.**
>
> ...and uses the **ForkJoin** method to run these two tasks in parallel and return: **IObservable<TaskCompletedSummary[]>.**

...and uses **ObserveOnDispatcher** method to update the UI with the information in the **TaskCompletedSummary[]** array.

FIGURE 5
The synchronization methods

The **StartSynchronization** method in the **SurveysSynchronizationService** class uses the **Observable.ForkJoin** method to define the set of parallel operations that make up the synchronization process. The **ForkJoin** method blocks until all the parallel operations are complete.

The following code example shows the **SurveysSynchronizationService** class, from the TailSpin.PhoneServices project, and includes an outline of the **StartSynchronization** method that the **SurveyListViewModel** class calls. This code implements the set of tasks shown in Figure 5.

C#

```
...
using Microsoft.Phone.Reactive;
...

public class SurveysSynchronizationService :
  ISurveysSynchronizationService
{
  ...
```

```
  public IObservable<TaskCompletedSummary[]> StartSynchronization()
  {
    var surveyStore = this.surveyStoreLocator.GetStore();

    var getNewSurveys = GetNewSurveys(surveyStore);
    var saveSurveyAnswers = UploadSurveys(surveysStore);

    return Observable.ForkJoin(getNewSurveys, saveSurveyAnswers);
  }
}
```

*The application uses the Funq dependency injection container to create the **SurveysSynchronization-Service** instance. For more information, see the **ViewModelLocator** class.*

The StartSynchronization method uses Rx to run the two synchronization tasks asynchronously and in parallel. When each task completes, it returns a summary of what happened in a TaskCompleted-Summary object, and when both tasks are complete, the method returns an array of TaskCompleted-Summary objects from the ForkJoin method.

The getNewSurveys Task

The **getNewSurveys** task retrieves a list of new surveys from the Tailspin Surveys service and saves them in isolated storage. When the task is complete, it creates a **TaskCompletedSummary** object with information about the task. The following code example shows the partial definition of this task that breaks down to the following subtasks:

- The **GetNewSurveys** method returns an observable list of **SurveyTemplate** objects from the Tailspin Surveys service.
- The **Select** method saves these surveys to isolated storage on the phone, updates the last synchronization date, and then returns an observable **TaskCompletedSummary** object.
- The Catch method traps any **WebException** and **UnauthorizedAccessException** errors and returns a **TaskCompletedSummary** object with details of the error.

C#

```
var getNewSurveys =
  this.surveysServiceClientFactory()
  .GetNewSurveys(surveyStore.LastSyncDate)
  .Select(surveys =>
  {
    surveyStore.SaveSurveyTemplates(surveys);

    if (surveys.Count() > 0)
    {
      surveyStore.LastSyncDate = surveys.Max(s => s.CreatedOn).ToString("s");
    }

    ...
```

```
    return new TaskCompletedSummary
      {
        Task = GetSurveysTask,
        Result = TaskSummaryResult.Success,
        Context = surveys.Count().ToString()
      };
})
.Catch(
(Exception exception) =>
{
  if (exception is WebException)
  {
    var webException = exception as WebException;
    var summary = ExceptionHandling.GetSummaryFromWebException(
      GetSurveysTask, webException);
    return Observable.Return(summary);
  }

  if (exception is UnauthorizedAccessException)
  {
    return Observable.Return(new TaskCompletedSummary
    {
      Task = GetSurveysTask,
      Result = TaskSummaryResult.AccessDenied
    });
  }

  throw exception;
});
```

The saveSurveyAnswers Task

The **saveSurveyAnswers** task saves completed survey answers to the Tailspin Surveys service and then removes them from isolated storage to free up storage space on the phone. It returns an observable **TaskCompletedSummary** object with information about the task. The following code example shows the complete definition of this task that breaks down to the following subtasks:

1. The **GetCompleteSurveyAnswers** method from the **SurveyStore** class gets a list of completed surveys from isolated storage.
2. The first call to **Observable.Return** creates an observable **TaskCompletedSummary** object so that the task returns at least one observable object (otherwise, the **ForkJoin** method may never complete). This also provides a default value to return if there are no survey answers to send to the Tailspin Surveys service.
3. The **SaveSurveyAnswers** method from the **SurveysServiceClient** class saves the completed surveys to the Tailspin Surveys service and returns **IObservable<Unit>** indicating whether the operation was successful or not.

4. The **Select** method deletes all the completed surveys from isolated storage and then returns an observable **TaskCompletedSummary** object.
5. The **Catch** method traps any **WebException** and **UnauthorizedAccessException** errors and returns a **TaskCompletedSummary** object with details of the error.

```
C#
var surveyAnswers = surveyStore.GetCompleteSurveyAnswers();
var saveSurveyAnswers = Observable.Return(
  new TaskCompletedSummary
{
  Task = SaveSurveyAnswersTask,
  Result = TaskSummaryResult.Success,
  Context = 0.ToString()
});

if (surveyAnswers.Count() > 0)
{
  saveSurveyAnswers =
  this.surveysServiceClientFactory()
    .SaveSurveyAnswers(surveyAnswers)
    .Select(unit =>
    {
      var sentAnswersCount = surveyAnswers.Count();
      surveyStore.DeleteSurveyAnswers(surveyAnswers);
      return new TaskCompletedSummary
      {
        Task = SaveSurveyAnswersTask,
        Result = TaskSummaryResult.Success,
        Context = sentAnswersCount.ToString()
      };
    })
    .Catch(
    (Exception exception) =>
    {
      ...
    });
}
```

You can have secondary Tiles for an application on Start without having an Application Tile.

Using Live Tiles on the Phone

The Tailspin mobile client must be able to support pinning Tiles to Start. This section describes how Tailspin designed and implemented this functionality.

A Tile is a link to an application displayed in Start. There are two types of Tiles:

- The Application Tile is the Tile created when a user pins an application to Start. Tapping a pinned Application Tile navigates the user to the application's opening page.
- A secondary Tile is created programmatically by an application based on an interaction from the user. A typical use for a secondary Tile is to pin a page other than the homepage to Start, for quick access. The application's code provides the Tile with launch parameters to customize the navigation destination of the Tile. For example, tapping on a secondary Tile that represents a survey would open the survey in the Tailspin mobile client application.

An Application Tile can be created by the user only when the user taps and holds the application name in the Application List and then selects **pin to start**. Therefore, when an Application Tile is created, the application is already deactivated. Secondary Tiles can be created only as a result of user input in the application, and when a secondary Tile is created, the application is deactivated. For more information see, *"Tiles Overview for Windows Phone,"* on MSDN.

Tiles are two-sided and display information by flipping between the front and back sides of the Tile.

The elements on the front of a Tile are:

- A background PNG or JPG image for the Tile that should be 173 pixels wide by 173 pixels high
- A title string that overlays the background image
- A count value that overlays the background image; for example, the number of new surveys available

The elements on the back of a Tile are:

- A background PNG or JPG image for the Tile that should be 173 pixels wide by 173 pixels high
- A title string that overlays the background image at the bottom of the Tile
- A content string that overlays the background image in the body of the back of the Tile

If none of the properties on the back side of the Tile are set, the Tile will not flip over and only the front side of the Tile will be displayed. For guidelines about how to design a Tile, see the "Start Tiles" section in *"Essential Graphics, Visual Indicators, and Notifications for Windows Phone"* on MSDN.

Overview of the Solution

There are two separate pinning scenarios that Tailspin wanted the mobile client to support:

- The mobile client must be capable of pinning an Application Tile to Start that includes a count of the number of new surveys that have been downloaded since the application was last opened. Tapping the tile should launch the application. In addition, when the application is deactivated, the count value on the Application Tile should be reset.
- The mobile client must be capable of pinning surveys to secondary Tiles on Start. Each secondary Tile should contain a title string that represents the survey name. Tapping the tile should launch the application and navigate the user to the page containing the survey.

Based upon these scenarios, it was not necessary to use two-sided tiles.

Inside the Implementation

Now is a good time to walk through the code that implements live tiles in more detail. As you go through this section, you may want to download the *Windows Phone Tailspin Surveys* application from the Microsoft Download Center.

The Application Tile

The **RunPeriodicTask** method of the **ScheduledAgent** class calls the **GetNewSurveys** method of the **SurveysSynchronization** class. The **GetNewSurveys** method downloads new surveys from the Tailspin cloud service through the background agent. In addition, it is responsible for updating the Application Tile **Count** property that represents the number of new surveys that have been downloaded since the application was last opened. The following code example shows the **GetNewSurveys** method.

```
C#
public IObservable<TaskCompletedSummary>
  GetNewSurveys(ISurveyStore surveyStore)
{
  var getNewSurveys =
    surveysServiceClientFactory()
    .GetNewSurveys(surveyStore.LastSyncDate)
    .Select(surveys =>
    {
      surveyStore.SaveSurveyTemplates(surveys);

      ...

      var tile = ShellTile.ActiveTiles.First();
      var tileData = new StandardTileData()
      {
        Count = this.UnopenedSurveyCount
      };
      tile.Update(tileData);

      ...
    })
    ...

    return getNewSurveys;
}
```

The method retrieves the Tile for the application and then updates the **Count** property of the Tile to the value of the **UnopenedSurveyCount** property, which is the number of surveys that have been downloaded since the mobile client application was deactivated, before updating the Tile on Start.

The **UnopenedSurveyCount** property simply retrieves the value of the **UnopenedCount** property in the **SurveysList** class. The **GetNewSurveys** method calls the **SaveSurveyTemplates** method to save the newly downloaded surveys and the **SaveSurveyTemplates** method updates the **UnopenedCount** property. The following code example shows the **SaveSurveyTemplates** method.

C#
```
public void SaveSurveyTemplates(IEnumerable<SurveyTemplate> surveys)
{
  foreach (var s in surveys)
  {
    s.IsNew = true;
  }

  var newSurveys = surveys.Where(ns => !AllSurveys.Templates.Any(
    s => s.SlugName == ns.SlugName && s.Tenant == ns.Tenant));

  AllSurveys.UnopenedCount += newSurveys.Count();
  SaveUnopenedCount();

  AllSurveys.Templates.AddRange(newSurveys);
  SaveTemplates();
}
```

The method marks each downloaded survey as new, through the **SurveyTemplate.IsNew** property. It then builds a collection of new surveys before incrementing the **UnopenedCount** property of the AllSurveys collection by the number of surveys in the newSurveys collection. The call to the **SaveUnopenedCount** method saves the value of the **UnopenedCount** property to isolated storage.

As has been previously mentioned, the count value on the Application Tile represents the number of new surveys that have been downloaded since the application was last opened. When the mobile client application is closed by the user navigating backwards past the first page of the application, the **ResetUnopenedSurveyCount** method of the **SurveyListViewModel** is called, which resets the count value on the Application Tile. The following code example shows the **ResetUnopenedSurveyCount** method.

C#
```
public void ResetUnopenedSurveyCount()
{
  synchronizationService.UnopenedSurveyCount = 0;
  var tile = shellTile.ActiveTiles.First();
  var tileData = new StandardTileData()
  {
    Count = 0
  };
  tile.Update(tileData);
}
```

The method resets the **UnopenedSurveyCount** property of the **SurveysSynchronizationService** class to 0, which updates both the AllSurveys collection and isolated storage. The method then retrieves the Application Tile for the application and resets the **Count** property of the Tile to 0, before updating the Tile on Start.

Secondary Tiles

Surveys can be pinned to Start as secondary Tiles from two locations in the mobile client application.

- From a context **MenuItem** on each survey in the SurveyListView. This binds to the **PinCommand** property in the **SurveyTemplateViewModel** class.
- From an **ApplicationBarButtonCommand** in the TakeSurveyView. This binds to the **PinCommand** property in the **TakeSurveyViewModel**.

The following code example shows the initialization of the **PinCommand** property from the **SurveyTemplateViewModel** class, and the **Actions** executed by it.

C#

```
private readonly IShellTile shellTile;
public DelegateCommand PinCommand { get; set; }
...

public SurveyTemplateViewModel(
    SurveyTemplate surveyTemplate,
    INavigationService navigationService,
    IPhoneApplicationServiceFacade phoneApplicationServiceFacade,
    IShellTile shellTile,
    ILocationService locationService)
{
  ...
  this.shellTile = shellTile;
  ...

  this.PinCommand = new DelegateCommand(PinToStart, () => IsPinnable);
  ...
}

public bool IsPinnable
{
  get
  {
    return shellTile.ActiveTiles.FirstOrDefault(
      x => x.NavigationUri.ToString() ==
     string.Format(Constants.PinnedSurveyUriFormat,
     this.Template.SlugName)) == null;
  }
}

public void PinToStart()
{
  var tile = shellTile.ActiveTiles.FirstOrDefault(
    x => x.NavigationUri.ToString() ==
    string.Format(Constants.PinnedSurveyUriFormat,
    this.Template.SlugName));
```

```
  if (tile == null)
  {
    var tileData = new StandardTileData
    {
      Title = this.Template.Title,
      BackgroundImage = new Uri("/Background.png", UriKind.Relative)
    };

    shellTile.Create(new Uri(string.Format(
      Constants.PinnedSurveyUriFormat, this.Template.SlugName),
      UriKind.Relative), tileData);
  }
}
```

In the **SurveyTemplateViewModel** constructor the **PinCommand** is initialized so that its execute Action is set to the **PinToStart** method, and it can execute **Action** is set to the **IsPinnable** property.

The **IsPinnable** property checks to see if the survey is already pinned to Start by comparing the **NavigationUri** of each secondary Tile to a string that will be the **NavigationUri** of the survey. If the two match, the property is set to false and the user will not be given the option to pin the survey.

The **PinToStart** method checks if the survey is already pinned to Start by comparing the **NavigationUri** of each secondary Tile to a string that will be the **NavigationUri** of the survey. If the survey is not present on Start, a new secondary Tile is pinned to Start by a call to **shellTile.Create**. The Tile data includes a title string and a background image. A launch parameter is also specified in order to specify that the navigation destination of the Tile is the survey represented by its **SlugName**.

When the user attempts to pin a survey from the TakeSurveyView, the PinCommand in the TakeSurveyViewModel class examines the IsPinnable property in the SurveyTemplateViewModel class to see if the survey can be pinned and if it can be, the PinToStart method in the SurveyTemplateViewModel class is executed.

The following code example shows the OnNavigatedTo method in the TakeSurveyView class. When the user taps on a secondary Tile, this method is called.

C#

```
protected override void OnNavigatedTo(NavigationEventArgs e)
{
  if (((TakeSurveyViewModel)this.DataContext).TemplateViewModel == null)
  {
    string surveyId;
    if (NavigationContext.QueryString.ContainsKey("surveyID"))
    {
      surveyId = NavigationContext.QueryString["surveyID"];
    }
    else
    {
      surveyId = (string)PhoneApplicationService.Current.State["TakeSurveyId"];
    }
```

```
        ((TakeSurveyViewModel)this.DataContext).Initialize(surveyId);
    }
}
```

This method checks whether the page is being navigated to from the secondary Tile by checking if the **TemplateViewModel** is null. If it is, it checks if the **QueryString** contains a surveyID and retrieves the surveyID from the **QueryString** if this is the case. Otherwise, it retrieves the surveyID from the application state dictionary. It then initializes the **DataContext** by calling an overload of the **Initialize** method that takes the surveyID as a parameter. The outcome is that the specified survey is loaded and navigated to.

Higher accuracy in location data requires higher power consumption.

Using Location Services on the Phone

Tailspin would like to capture users locations when they are answering a survey and include this location information as part of the survey data that's sent to the Tailspin service when the synchronization process runs. Tenants can use the location information when they analyze the survey results.

The Tailspin Surveys website displays survey location data using a Bing® search engine maps control. For more information, see the SuveyLocation.ascx file in the TailSpin.Web project. There is also a Bing maps control available for Windows Phone.

You could also create a dummy implementation of the **ILocationService** interface that returns a fixed location to use in the emulator. Alternatively, you can simulate location data using the location sensor simulator in the emulator.

Overview of the Solution

The Windows Phone API includes a Location Service that wraps the available hardware on the phone and enables your application to easily access location data. However, there is a trade-off between the accuracy of the data you can obtain from the Location Service and your application's power consumption. Tailspin does not require highly accurate location data for their surveys, so they have optimized the Surveys mobile client application to minimize power consumption.

The developers at Tailspin also decided that it is more important to save the survey data quickly and reliably, and not wait if the location data is not currently available. It can take up to 120 seconds to get location data back from the Location Service. Therefore, if the Location Service doesn't have the location data when the Tailspin mobile client application requests it, the latest available location data is saved along with the survey answers, instead of waiting for new data. Furthermore, the application only asks the Location Service for location data when the Surveys application needs it to save with a survey, although in some applications, you might consider caching the available location data at fixed intervals.

The sample application asks the user's permission to collect location data in the AppSettingsView page. In your own application, you must obtain the user's consent before collecting and using location data, typically on an initial settings screen when the application first runs. You should also make sure that your application can continue to function if the user disables the Location Service or doesn't give their consent for your application to use the phone's location data.

Inside the Implementation

Now is a good time to walk through the code that acquires location data from the phone in more detail. As you go through this section, you may want to download the *Windows Phone Tailspin Surveys* application from the Microsoft Download Center.

The following code example shows the **ILocationService** interface from the TailSpin.PhoneServices project.

```
C#
public interface ILocationService
{
  GeoCoordinate TryToGetCurrentLocation();
  void StartWatcher();
  void StopWatcher();
}
```

The **TryToGetCurrentLocation** method returns a **GeoCoordinate** object that holds the location data. The **LocationService** class implements this interface.

The LocationService class uses the GeoCoordinateWatcherAdapter class from the TailSpin.Phone.Adapters project to retrieve the current location from the phone when the user starts a survey and when the user completes a survey. The GeoCoordinateWatcherAdapter class implements the IGeoCoordinateWatcher interface and adapts the GeoCoordinateWatcher class from the Windows Phone API in order to create a loosely coupled class that is testable. Tailspin does not require highly accurate location data in the Surveys application, so the ContainerLocator class initializes the GeoCoordinateWatcherAdapter class using the default accuracy. This gives the phone the opportunity to reduce its power consumption and to return location data more quickly. The following code example shows the initialization of the GeoCoordinateWatcherAdapter class in the ContainerLocator class.

C#

```
private void ConfigureContainer()
{
  ...
  this.Container.Register<IGeoCoordinateWatcher>(c =>
    new GeoCoordinateWatcherAdapter(GeoPositionAccuracy.Default));
  ...
}
```

The **StatusChanged** event in **GeoCoordinateWatcherAdapter** class indicates that the status of the **GeoCoordinateWatcherAdapter** object has changed. The status could be Ready, Initializing, NoData, or Disabled. The **GeoCoordinateWatcherAdapter** class uses the **StatusChanged** event to indicate the ability of the location provider to provide location updates. The following code example shows the **TryToGetCurrentLocation**, **StartWatcher**, and **StopWatcher** methods from the **LocationService** class.

C#

```
private readonly TimeSpan maximumAge = TimeSpan.FromMinutes(15);
private GeoCoordinate lastCoordinate = GeoCoordinate.Unknown;
private DateTime lastCoordinateTime;
private IGeoCoordinateWatcher geoCoordinateWatcher;
...

public GeoCoordinate TryToGetCurrentLocation()
{
  if (!settingsStore.LocationServiceAllowed)
  {
    return GeoCoordinate.Unknown;
  }

  if (geoCoordinateWatcher.Status == GeoPositionStatus.Ready)
  {
    lastCoordinate = geoCoordinateWatcher.Position.Location;
    lastCoordinateTime = geoCoordinateWatcher.Position.Timestamp.DateTime;
    return lastCoordinate;
  }

  if (maximumAge < (DateTime.Now - lastCoordinateTime))
  {
    return GeoCoordinate.Unknown;
  }
  else
  {
    return lastCoordinate;
  }
}
```

```
public void StartWatcher()
{
  this.geoCoordinateWatcher.Start();
}

public void StopWatcher()
{
  this.geoCoordinateWatcher.Stop();
}
```

The **TryToGetCurrentLocation** method only returns location data if the user has given consent for Tailspin to use location data obtained from the phone; otherwise, **GeoCoordinate.Unknown** is returned. If the user has given consent for Tailspin to use location data, the location and the time the location was acquired are obtained from the GeoCoordinateWatcherAdapter class, provided that the Status property of the **GeoCoordinateWatcherAdapter** class is **GeoPositionStatus.Ready**. If the Status property of the **GeoCoordinateWatcherAdapter** class is not **GeoPositionStatus.Ready**, the method returns the last available location data, provided that it was obtained within the last 15 minutes. Otherwise, it returns **GeoCoordinate.Unknown**.

The **StartWatcher** and **StopWatcher** methods are used by the **TakeSurveyViewModel** class to initiate and stop the acquisition of data from the current location provider, respectively.

Acquiring Image and Audio Data on the Phone

The Tailspin Surveys mobile client application allows users to capture images from the device's camera and record audio from the device's microphone as answers to survey questions. The application saves the captured data as part of the survey answer, and this data is sent to the Tailspin Surveys web service when the user synchronizes the mobile client application.

The mobile client application can capture image and audio data.

Overview of the Solution

The techniques you use to capture audio and image data are different. To capture image data from the camera, you use a "chooser," and to capture audio data from the microphone, you must use interop with the XNA® development platform framework on the phone.

Capturing Image Data

The chooser for capturing image data is the **CameraCaptureTask**. When you use a chooser, the operating system deactivates your application and runs the chooser as a new process. When the chooser has completed its task, the operating system reactivates your application and delivers any data to your application using a callback method. The developers at Tailspin chose to implement this using the **Observable** class from the Rx library on the phone. The Tailspin Surveys mobile client application saves the captured picture to isolated storage along with the other survey answers and also displays the picture in the view.

Recording Audio Data

To access the microphone on the Windows Phone device, you have to use XNA; it's not possible to access it directly from Silverlight. To interoperate with XNA, you must use an XNA asynchronous event dispatcher to connect to the XNA events from Silverlight. Your application can then handle the microphone events that deliver the raw audio data to your application. Your application must then convert or encode the audio data to a valid sound format before saving it in isolated storage.

Inside the Implementation

Now is a good time to take a more detailed look at the code that captures image data and records audio.

As you go through this section, you may want to download the *Windows Phone Tailspin Surveys* application from the Microsoft Download Center.

Capturing Image Data

The **ICameraCaptureTask** interface in the TailSpin.PhoneClient.Adapters project defines signatures for a property, an event, and a method. The following code example shows this interface.

```
C#
public interface ICameraCaptureTask
{
  SettablePhotoResult TaskEventArgs
  {
      get;
      set;
  }

  event EventHandler<SettablePhotoResult> Completed;

  void Show();
}
```

The **ICameraCaptureTask** interface is implemented by the **CameraCaptureTaskAdapter** class, which adapts the **CameraCaptureTask** class from the API. The purpose of adapting the **CameraCapture-Task** class with a class that implements **ICameraCaptureTask** is to create a loosely coupled class that is testable. The following code example shows the class.

```
C#
public class CameraCaptureTaskAdapter : ICameraCaptureTask
{
  public CameraCaptureTaskAdapter()
  {
      WrappedSubject = new CameraCaptureTask();
      ...
  }

  private CameraCaptureTask WrappedSubject { get; set; }
```

```
  public SettablePhotoResult TaskEventArgs
  {
    get
    {
      return new SettablePhotoResult(WrappedSubject.TaskEventArgs);
    }
    set
    {
      WrappedSubject.TaskEventArgs = value;
    }
  }

  public event EventHandler<SettablePhotoResult> Completed;

  ...

  public void Show()
  {
    WrappedSubject.Show();
  }

  ...
}
```

The **SettablePhotoResult** class provides an implementation of the **PhotoResult** class where the **ChosenPhoto** and **OriginalFileName** properties are settable. The following code example shows the class.

C#

```
public class SettablePhotoResult : PhotoResult
{
  public SettablePhotoResult(PhotoResult photoResult)
  {
    ChosenPhoto = photoResult.ChosenPhoto;
    OriginalFileName = photoResult.OriginalFileName;
    Error = photoResult.Error;
  }

  public SettablePhotoResult()
  {
  }

  public new Stream ChosenPhoto { get; set; }
  public new string OriginalFileName { get; set; }
  public new Exception Error { get; set; }
}
```

When **TakeSurveyViewModel** creates an instance of the **PictureQuestionViewModel** class, it passes in a new instance of the **CameraCaptureTaskAdapter** class, which in turn creates an instance of the **CameraCaptureTask** class.

The following code example shows how the **CameraCaptureCommand** delegate command is defined in the constructor for the **PictureQuestionViewModel** class. This command uses the **Capturing** property to check whether the application is already in the process of capturing a picture and to control whether the command can be invoked from the UI. The method displays a picture if there is already one saved for this question.

The method then uses the CameraCaptureTaskAdapter instance, which will launch the chooser for taking the photograph and return the captured picture.

You can use the Exchangeable Image File (EXIF) data in the picture to determine the correct orientation for displaying the picture.

C#
```
private readonly ICameraCaptureTask task;
...

public PictureQuestionViewModel(QuestionAnswer questionAnswer,
  ICameraCaptureTask cameraCaptureTask, IMessageBox messageBox)
  : base(questionAnswer)
{
  this.CameraCaptureCommand =
    new DelegateCommand(this.CapturePicture,
      () => !this.Capturing);
  if (questionAnswer.Value != null)
  {
    this.LoadPictureBitmap(questionAnswer.Value);
  }
  this.task = cameraCaptureTask;
  ...
}

public DelegateCommand CameraCaptureCommand { get; set; }
...

public bool Capturing
{
  get { return this.capturing; }
  set
  {
    if (this.capturing != value)
    {
      this.capturing = value;
      this.RaisePropertyChanged(() => this.Capturing);
```

```
    }
  }
}

...

private void CapturePicture()
{
  if (!this.Capturing)
  {
    this.task.Show();
    this.Capturing = true;
    this.CameraCaptureCommand.RaiseCanExecuteChanged();
  }
}
```

The following code examples show how the constructor uses the **Observable.FromEvent** method to specify how to handle the **Completed** event raised by the **CameraCaptureTask** chooser object when the user has finished with the chooser. The first example shows how the application saves the picture if the **CameraCaptureTask** succeeds.

C#

```
Observable.FromEvent<SettablePhotoResult>(
  h => this.task.Completed += h,
  h => this.task.Completed -= h)
  .Where(e => e.EventArgs.ChosenPhoto != null)
  .Subscribe(result =>
  {
    this.Capturing = false;
    SavePictureFile(result.EventArgs.ChosenPhoto);
    this.Answer.Value = this.fileName;
    this.RaisePropertyChanged(string.Empty);
    this.CameraCaptureCommand.RaiseCanExecuteChanged();
  });
```

The second example shows how the **CameraCaptureCommand** command is re-enabled if the user didn't take a picture.

C#

```
Observable.FromEvent<SettablePhotoResult>(
  h => this.task.Completed += h,
  h => this.task.Completed -= h)
  .Where(e => e.EventArgs.ChosenPhoto == null &&
    e.EventArgs.Error == null)
  .Subscribe(p =>
  {
    this.Capturing = false;
    this.CameraCaptureCommand.RaiseCanExecuteChanged();
  });
```

The third example shows how a message box containing an error message is displayed to the user, if the **CameraCaptureTask** fails.

C#
```
Observable.FromEvent<SettablePhotoResult>(
  h => this.task.Completed += h,
  h => this.task.Completed -= h)
  .Where(e => e.EventArgs.Error != null &&
    !string.IsNullOrEmpty(e.EventArgs.Error.Message))
  .Subscribe(p =>
  {
    this.Capturing = false;
    this.messageBox.Show(p.EventArgs.Error.Message);
    this.CameraCaptureCommand.RaiseCanExecuteChanged();
  });
```

The **SavePictureFile** method uses an efficient approach to persisting the captured image to isolated storage, which avoids writing a single byte of the image to the file during each iteration of a loop. This has the additional advantage of not requiring a progress indicator to inform the user of progress through the save process. The following code example shows the method.

C#
```
private void SavePictureFile(Stream chosenPhoto)
{
  SavingPictureFile = true;

  // Store the image bytes.
  byte[] imageBytes = new byte[chosenPhoto.Length];
  chosenPhoto.Read(imageBytes, 0, imageBytes.Length);

  // Seek back so we can create an image.
  chosenPhoto.Seek(0, SeekOrigin.Begin);

  // Create an image from the stream.
  var imageSource = PictureDecoder.DecodeJpeg(chosenPhoto);
  this.Picture = imageSource;

  // Save the stream
  var isoFile = IsolatedStorageFile.GetUserStoreForApplication();
  using (var stream = isoFile.CreateFile(filename))
  {
    stream.Write(imageBytes, 0, imageBytes.Length);
  }

  SavingPictureFile = false;
}
```

If your application is camera intensive, you should consider using the PhotoCamera API class, which allows you to configure functionality such as image capture, focus, resolution, and flash mode.

The drawback with this method is that the **CameraCaptureTask** class returns a high resolution image, which the **SavePictureFile** method simply writes to a file. The **PictureQuestionViewModel** also displays the captured image. However, since high-resolution images consume a lot of memory, the mobile application client crashes when a survey contains too many picture questions containing captured images. An alternative approach would be to use the **CameraCaptureTask** class for image capture, and then reduce the resolution of the captured image before consuming it. Another alternative approach would to be to use the **PhotoCamera** API class for image capture, and configure the capture resolution prior to capturing images.

Using XNA Interop to Record Audio

Before the Tailspin mobile client application can handle events raised by XNA objects, such as a Microphone object, it must create an XNA asynchronous dispatcher service. The following code example from the VoiceQuestionViewModel class shows how this is done.

```
C#
...
private XnaAsyncDispatcher xnaAsyncDispatcher;
...

public VoiceQuestionViewModel(QuestionAnswer questionAnswer,
  IIsolatedStorageFacade isolatedStorageFacade,
  INavigationService navigationService)
    : base(questionAnswer)
{
  ...
  xnaAsyncDispatcher = new XnaAsyncDispatcher(
    TimeSpan.FromMilliseconds(50)));
  xnaAsyncDispatcher.StartService();
  ...
}
```

The VoiceQuestionView.xaml file defines two buttons, one toggles recording on and off, and the other plays back any saved audio. The recording toggle button is bound to the **DefaultActionCommand** command in the view model, and the play button is bound to the **PlayCommand** command in the view model.

The **DefaultAction** command uses the **StartRecording** and **StopRecording** methods in the **VoiceQuestionViewModel** class to start and stop audio recording. The following code example shows the StartRecording method.

```
C#
private MediaState priorMediaState;
...

private void StartRecording()
{
  StopPlayback();

  priorMediaState = MediaPlayer.State;
  if (priorMediaState == MediaState.Playing)
  {
    MediaPlayer.Pause();
  }

  var mic = Microphone.Default;
  if (mic.State == MicrophoneState.Started)
  {
    mic.Stop();
  }

  this.stream = new MemoryStream();

  buffer = new byte[mic.GetSampleSizeInBytes(mic.BufferDuration)];

  this.observableMic = Observable.FromEvent<EventArgs>(
    h => mic.BufferReady += h, h => mic.BufferReady -= h)
    .Subscribe(p => this.CaptureMicrophoneBufferResults());
  mic.Start();
}
```

This method determines if a media item is being played, and if it is, the media item is paused. The method then gets a reference to the default microphone on the device and creates a **MemoryStream** instance to store the raw audio data.

*You can find the **Microphone** class in the **Microsoft.Xna.Framework.Audio** namespace.*

The method uses the **Observable.FromEvent** method to subscribe to the microphone's **BufferReady** event, and whenever the event is raised, the application calls **CaptureMicrophoneBufferResults** to capture the raw audio data. Finally, the method starts the microphone.

The following code example shows the **StopRecording** method. It creates a **WaveFormatter** instance to convert the raw audio data to the WAV format before using the instance to write the audio data to isolated storage. It then disposes of the **MemoryStream, Microphone**, and **WaveFormatter** instances and attaches the name of the saved audio files to the question. Finally, if a media item was playing and was paused when the **StartRecording** method was called, the method resumes playback of the media item.

The Windows Phone API does not include any methods to convert audio formats. You can find the **WaveFormatter** class in the **TailsSpin.PhoneClient.Infrastructure** namespace.

C#
```
private MediaState priorMediaState;
...

private void StopRecording()
{
  var mic = Microphone.Default;
  this.CaptureMicrophoneBufferResults();
  mic.Stop();

  this.formatter = new WaveFormatter(this.wavFileName,
    (ushort)mic.SampleRate, 16, 1, isolatedStorageFacade);
  this.formatter.WriteDataChunk(stream.ToArray());

  this.stream.Dispose();
  this.observableMic.Dispose();
  this.formatter.Dispose();
  this.formatter = null;
  this.buffer = null;
  this.Answer.Value = this.wavFileName;

  if (priorMediaState == MediaState.Playing)
  {
    MediaPlayer.Resume();
  }
}
```

The play button in the VoiceQuestionView view plays the recorded audio by using the **SoundEffect** class from the **Microsoft.Xna.Framework.Audio** namespace. The following code example shows the **Play** method from the **VoiceQuestionViewModel** class that loads audio data from isolated storage and plays it back. Before loading and playing audio data from isolated storage, the method determines if a media item is already being played, and if it is, the media item is paused. Once playback of the audio data has finished, the method resumes playback of the original media item.

C#
```
private MediaState priorMediaState;
...

private void Play()
{
  priorMediaState = MediaPlayer.State;
  if (priorMediaState == MediaState.Playing)
```

```
{
    MediaPlayer.Pause();
  }

  fileSystem = IsolatedStorageFile.GetUserStoreForApplication();

  fileStream = fileSystem.OpenFile(this.wavFileName, FileMode.Open,
    FileAccess.Read);

  try
  {
    soundEffect = SoundEffect.FromStream(fileStream);
    soundEffectInstance = soundEffect.CreateInstance();
    soundEffectInstance.Play();
  }
  catch (ArgumentException)
  {
  }

  if (priorMediaState == MediaState.Playing)
  {
    MediaPlayer.Resume();
  }
}
```

Logging Errors and Diagnostic Information on the Phone

The sample application does not log any diagnostic information or details of error conditions from the Tailspin mobile client application. Tailspin plans to add this functionality to a future version of the mobile client application after they evaluate a number of tradeoffs.

For example, Tailspin must decide whether to keep a running log on the phone or simply report errors to a service as and when they occur. Keeping a log on the phone will use storage, so Tailspin would have to implement a mechanism to manage the amount of isolated storage used for log data, perhaps by keeping rolling logs of recent activity or implementing a purge policy. Sending error data as it occurs minimizes the storage requirements, but it makes it harder to access data about the state of the application before the error occurred. Tailspin would also need to develop a robust way to send the error information to a service, while transferring log files could take place during the application's standard synchronization process. Collecting logs also makes it easier to correlate activities on the phone with activities in the various Tailspin Surveys services.

Tailspin must also consider the user experience. A privacy policy would require a user to opt into collecting diagnostic information, and Tailspin might want to include options that would enable users to set the level of logging detail to collect.

Conclusion

This chapter described how the developers at Tailspin implemented the model elements from the MVVM pattern in the Tailspin Surveys mobile client application, and how the application leverages services offered by the Windows Phone platform, such as isolated storage and location services.

The developers at Tailspin also created some services themselves; for example, they created the synchronization service that runs both in the background, and in the foreground as a set of asynchronous parallel tasks that manage the data used by the application. This synchronization service needs to access data held remotely by the Tailspin Surveys application that runs on the Windows Azure™ technology platform. The next chapter will describe how the mobile client application can access remote services like the one that provides access to the data held in the Windows Azure application.

Questions

1. The Data Protection API (DPAPI) can be used to encrypt and decrypt data in isolated storage. What does the DPAPI use as an encryption key?
   a. A user-generated private key.
   b. The user credentials.
   c. The phone credentials.
   d. The user and phone credentials.

2. What happens when your application is reactivated?
   a. You return to the first screen in your application.
   b. The operating system makes sure that the screen is displayed as it was when the application was deactivated.
   c. The operating system recreates the navigation stack within your application.
   d. The **Launching** event is raised.

3. What data should you save when you handle the deactivation request?
   a. State data required to rebuild the state of the last screen that was active before the application was deactivated.
   b. State data required to rebuild the state of previous screens that user had navigated through before the application was deactivated.
   c. Data that is normally persisted to isolated storage by the application at some point.
   d. The currently active screen.

4. Why does Tailspin use the Reactive Extensions (Rx) for .NET?
   a. To handle notifications from the Microsoft Push Notification Service.
   b. To handle UI events.
   c. To manage asynchronous tasks.
   d. To make the code that implements the asynchronous and parallel operations more compact and easier to understand.

5. What factors should you consider when you use location services on the phone?
    a. What level of accuracy your application requires for its geo-location data.
    b. Whether the device has a built-in Global Positioning System (GPS).
    c. How quickly you need to obtain the current location.
    d. Whether the user has consented to allowing your application to use the phone's GPS data.
6. Which factors constrain the use of a **ResourceIntensiveTask** agent?
    a. Resource-intensive agents do not run unless the Windows Phone device is connected to an external power source.
    b. Resource-intensive agents do not run unless the Windows Phone device has a network connection over Wi-Fi or through a connection to a PC.
    c. Resource-intensive agents do not run unless the Windows Phone device's battery power is greater than 90%.
    d. Resource-intensive agents do not run unless the Windows Phone device screen is locked.

More Information

For more information about isolated storage, see *"Local Data Storage for Windows Phone"* on MSDN.

For more information about handling deactivation, reactivation, and tombstoning, see *"Execution Model for Windows Phone"* on MSDN.

For more information about launchers and choosers, see *"Launchers and Choosers for Windows Phone"* on MSDN.

For more information about Reactive Extensions, see *"Reactive Extensions for .NET Framework Overview for Windows Phone"* and *"The Reactive Extensions (Rx)"* on MSDN.

For more information about location services, see *"Location for Windows Phone"* on MSDN.

These and all links in this book are accessible from the book's online bibliography. You can find the bibliography on MSDN at: ***http://msdn.microsoft.com/en-us/library/gg490786.aspx***.

4 Connecting with Services

This chapter describes the various ways that the Tailspin Surveys mobile client interacts with external services, both custom services created by Tailspin, and services offered by third-party companies. Connecting to external services from a mobile client introduced a set of challenges for the development team at Tailspin to meet in the design and implementation of the mobile client, and in the services hosted in Windows Azure™ technology platform. The mobile client application must do the following:

- It must operate reliably with variable network connectivity.
- It must minimize the use of network bandwidth (which may be costly).
- It must minimize its impact on the phone's battery life.

The online service components must do the following:

- They must offer an appropriate level of security.
- They must be easy to develop client applications against.
- They must support a range of client platforms.

The key areas of functionality in the Tailspin Surveys application that this chapter describes include authenticating with a web service from an application on the phone, pushing notifications to Windows® Phone devices, and transferring data between a Windows Phone device and a web service.

Authenticating with the Surveys Service

The Surveys service running in Windows Azure needs to know the identity of the user who is using the mobile client application on the Windows Phone device for two reasons. First, when the mobile client requests a list of surveys, the service determines the contents of the list based on the preferences stored in the user's profile. Second, when the mobile client submits a set of survey answers, the Surveys service needs to record who submitted the survey in order to be able to calculate any rewards due the user.

The Windows Azure-based Surveys service needs to identify the user using the mobile client application.

Goals and Requirements

Tailspin wants to be able to change the way it authenticates users without requiring major changes to the Surveys application.

Tailspin wants to externalize as much as possible of the authentication and authorization functionality from the main Surveys application. This will give Tailspin the flexibility to make changes to the way they handle authentication and authorization in the future without affecting the Surveys application itself. For example, Tailspin may want to enable users to identify themselves by using a Windows Live® ID.

It's also important to ensure that the mechanism the mobile client uses to authenticate is easy to implement on the Windows Phone platform and any other mobile platforms that Tailspin may support in the future.

Overview of the Solution

Figure 1 shows a high-level view of the approach adopted by Tailspin.

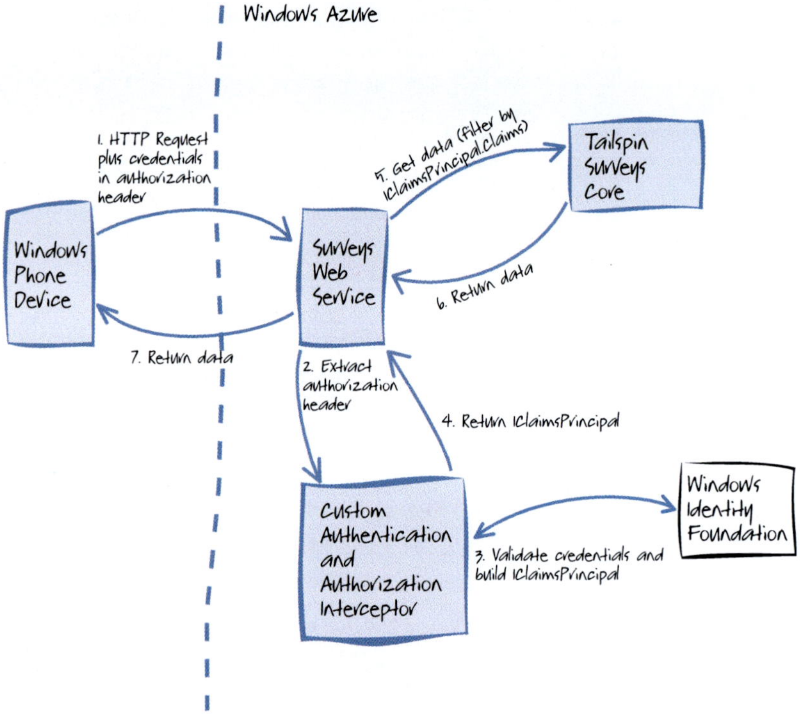

Figure 1
Authentication and authorization for the Surveys web services

The approach that Tailspin adopted assumes that the Windows Phone client application can send credentials to the Surveys web service in an HTTP header. The credentials could be a user name and password or a token. Tailspin could easily change the type of credentials in a future version of the application.

In the sample application, the phone sends a user name and password to demonstrate this approach. The mobile client does not perform any validation on the credentials that the user enters on the AppSettingsView page, but it does encrypt the password before it saves the credentials in isolated storage. In a real application, you may decide to enforce a password policy that requires strong passwords and regular password renewals.

In the Surveys web service, the custom authentication and authorization interceptor extracts the header that contains the user's credentials and identifies an authentication module to perform the authentication. It's possible that different client platforms use different authentication schemes, so the interceptor must be able to identify from the HTTP headers which type of authentication is being used, and then pass the credentials to the correct custom authentication module.

Tailspin must be able to handle authentication from other mobile platforms in the future, so choosing a flexible, standards-based approach to authentication is crucial.

The sample application uses a mock authentication module that simply checks for one of several hard-coded user names; it does not verify the password. If you use password-based credentials in your application, you should send a hashed version of the password over the network to compare with a hashed version stored in Windows Azure storage in your authentication module.

After the custom authentication module validates the credentials, it uses Windows Identity Foundation (WIF) to construct an **IClaimsPrincipal** object that contains the user's identity. In the future, this **IClaimsPrincipal** object might contain additional claims that the Surveys application could use to perform any authorization it requires.

The surveys web service includes the **IClaimsPrincipal.Claims.Name** value when it invokes any methods in the Tailspin Surveys core application. The Tailspin Surveys core application returns data for the user. In the future, the web service could also perform any necessary authorization before it invokes a method in the core application.

To change the authentication method for the mobile client application, Tailspin must make two changes to the application; they must:

1. Modify the mobile client to send the credentials in a custom HTTP header.
2. Add a new custom authentication module to validate the credentials and create an **IClaimsPrincipal** object.

When Tailspin developers were developing the mobile client, new versions of the OAuth and SWT standards were anticipated, so they decided to wait for these new releases.

A Future Claims-Based Approach

In the future, Tailspin is considering replacing the simple user name and password authentication scheme with a claims-based approach. One option is to use Simple Web Token (SWT) and the Open Authentication (OAuth) 2.0 protocol. This approach offers the following benefits:

- The authentication process is managed externally from the Tailspin Surveys application.
- The authentication process uses established standards.
- The Surveys application can use a claims-based approach to handle any future authorization requirements.

Figure 2 illustrates this approach, showing the external token issuer.

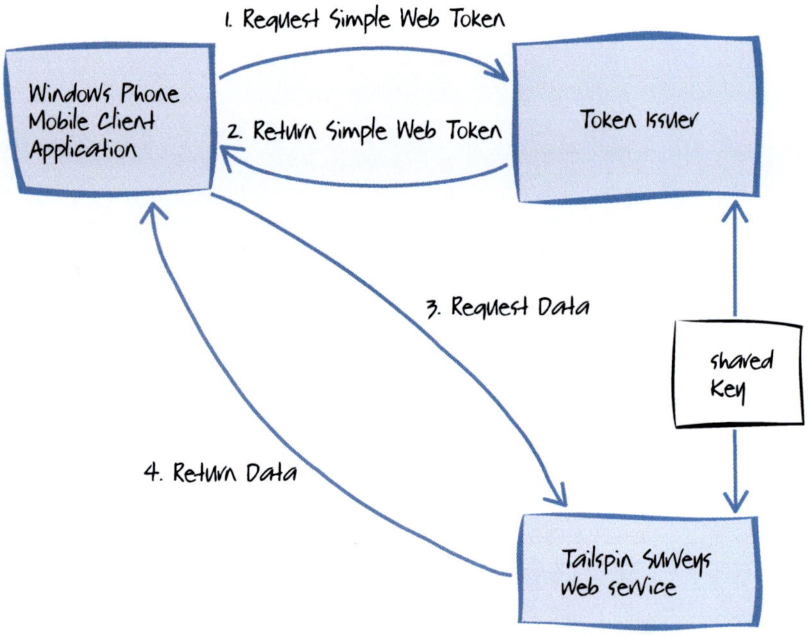

FIGURE 2
An authentication approach using SWT for the Surveys web service

In this scenario, before the mobile client application invokes a Surveys web service, it must obtain an SWT. It does this by sending a request to a token issuer that can issue SWTs; for example, Windows Azure Access Control Services (ACS). The request includes the items of information described in the following table.

Field	Description
Client ID	The client ID is an identifier for the consumer application, which is the Surveys service in this case.
Client Secret	A piece of information that proves that it is your application.
User Name	The user name of the person who wants to authenticate with the Surveys service. The application will prompt the user to enter this name in the user interface (UI).
Password	The user's password. The application will prompt the user to enter this password in the UI.

The client ID and client secret enable the issuer to determine which application is requesting an SWT. The issuer uses the user name and password to authenticate the user.

The token issuer then constructs an SWT containing the user's identity and any other claims that the consumer application (Tailspin Surveys) might require. The issuer also attaches a hash value generated using a secret key shared with the Tailspin Surveys service.

When the client application requests data from the Surveys service, it attaches the SWT to the request in the request's authorization header.

You should keep the Client ID and Client Secret secure on the phone. If someone discovers them, they could create an application that impersonates the Tailspin mobile client application.

When the Surveys service receives the request, a custom authentication module extracts the SWT from the authorization header, validates the SWT, and then extracts the claims from the SWT. The Surveys service can then use the claims with its authorization rules to determine what data, if any, it should return to the user.

The validation of the SWT in the custom authentication module performs the following steps.

- It verifies the hash of the SWT by using the shared secret key. This enables the Surveys service to verify the data integrity and the authenticity of the message.
- It verifies that the SWT has not expired. The token issuer sets the expiration time when it creates the SWT.
- It checks that the issuer that created the SWT is an issuer that the service is configured to trust.
- It checks that the client application that is making the request is a client that the service is configured to trust.

The OAuth protocol uses a shared key to generate a hash of the SWT. This shared secret key must be known by the issuer and the Surveys service.

Inside the Implementation

Now is a good time to walk through the code that implements the authentication process in more detail. As you go through this section, you may want to download the *Windows Phone Tailspin Surveys* application from the Microsoft Download Center.

The **CustomServiceHostFactory** class in the TailSpin.Services.Surveys project initializes the Surveys service. The following code example shows how this factory class creates the authorization manager.

C#
```
public class CustomServiceHostFactory : WebServiceHostFactory
{
  private readonly IUnityContainer container;

  public CustomServiceHostFactory(IUnityContainer container)
  {
    this.container = container;
  }

  protected override ServiceHost CreateServiceHost(
    Type serviceType, Uri[] baseAddresses)
  {
    var host = new CustomServiceHost(
      serviceType, baseAddresses, this.container);

    host.Authorization.ServiceAuthorizationManager =
      new SimulatedWebServiceAuthorizationManager();
    host.Authorization.PrincipalPermissionMode =
      PrincipalPermissionMode.Custom;

    return host;
  }
}
```

The sample Surveys application uses a simulated authorization manager. You must replace this with a real authorization manager in a production application.

The following code example from the **SimulatedWebServiceAuthorizationManager** class shows how to override the **CheckAccessCore** method in the **ServiceAuthorizationManager** class to provide a custom authorization decision.

C#
```
protected override bool CheckAccessCore(
  OperationContext operationContext)
{
  try
  {
    if (WebOperationContext.Current != null)
    {
      var headers =
        WebOperationContext.Current.IncomingRequest.Headers;
      if (headers != null)
      {
        var authorizationHeader =
          headers[HttpRequestHeader.Authorization];
        if (!string.IsNullOrEmpty(authorizationHeader))
        {
```

```
          if (authorizationHeader.StartsWith("user",
            StringComparison.OrdinalIgnoreCase))
          {
            var userRegex = new Regex(@"(\w+):([^\s]+)",
              RegexOptions.Singleline);
            var username = userRegex.Match(authorizationHeader)
              .Groups[1].Value;
            var password = userRegex.Match(authorizationHeader)
              .Groups[2].Value;
            if (ValidateUserAndPassword(username, password))
            {
              var identity = new ClaimsIdentity(new[]
              {
                new Claim(System.IdentityModel.Claims.ClaimTypes.Name,
                  username)
              }, "TailSpin");

              var principal =
                ClaimsPrincipal.CreateFromIdentity(identity);
              operationContext.ServiceSecurityContext
                .AuthorizationContext.Properties["Principal"] =
                principal;

              return true;
            }
          }
        }
      }
    }
  }
  catch (Exception)
  {
    if (WebOperationContext.Current != null)
    {
      WebOperationContext.Current.OutgoingResponse.StatusCode =
        HttpStatusCode.Unauthorized;
    }
    return false;
  }
  if (WebOperationContext.Current != null)
  {
    WebOperationContext.Current.OutgoingResponse.StatusCode =
      HttpStatusCode.Unauthorized;
  }
  return false;
}
```

In a real implementation, Tailspin could extract a token from the authorization header and use a validation routine that verifies the token.

The HttpWebRequest class provides an HTTP-specific implementation of the WebRequest class.

In this simulated authorization manager class, the **CheckAccess-Core** method extracts the user name and password from the authorization header, calls a validation routine, and if the validation routine succeeds, it attaches a **ClaimsPrincipal** object to the web service context.

In the sample application, the validation routine does nothing more than check that the user name is one of several hard-coded values.

The **IHttpWebRequest** interface, in the TailSpin.Phone.Adapters project, defines method signatures and properties that are implemented by the **HttpWebRequestAdapter** class. This class adapts the **HttpWebRequest** class from the API. The purpose of adapting the **HttpWebRequest** class with a class that implements **IHttpWeb-Request** is to create a loosely coupled class that is testable.

When **SurveysServiceClient** calls the **GetRequest** method in the **HttpClient** class, it passes in a new instance of **HttpWebRequest-Adapter**, which in turn creates an instance of **WebRequest**.

The following code example shows how the **GetRequest** method in the **HttpClient** class adds the authorization header with the user name and password credentials to the HTTP request that the mobile client sends to the various Tailspin web services.

C#
```
public IHttpWebRequest GetRequest(IHttpWebRequest httpWebRequest,
  string userName, string password)
{
  var authHeader = string.Format(CultureInfo.InvariantCulture,
    "user {0}:{1}", userName, password);
  httpWebRequest.Headers[HttpRequestHeader.Authorization] =
    authHeader;
  return httpWebRequest;
}
```

Notifying the Mobile Client of New Surveys

Tailspin wants a way to notify users of new surveys from the user's list of preferred tenants. Tenants are subscribers to the cloud-based

There are two types of notifications that the phone can receive when the application isn't running. The first is a toast notification that the phone displays in an overlay on the user's current screen. The user can click the message to launch the application. The second type of notification is a tile notification that changes the appearance of the front and back of the application's tile in the Quick Launch area of the phone's Start experience. If you change the appearance of a tile by sending a tile notification, and if you want to change the tile back to its original state, you'll have to send another message.
You can also use raw notifications to send data directly to the application, but this type of notification requires the application to be running in the foreground. If the application is not running, MPNS discards the notification and notifies the sender.

Tailspin Surveys application who publish surveys. Users will then be able to synchronize the mobile client application and start using the new surveys.

Overview of the Solution

Tailspin chose to use the Microsoft Push Notifications Service (MPNS) for Windows Phone to deliver information about relevant new surveys to users with Windows Phone devices. This feature allows the cloud-based Surveys application to notify a user about a relevant new survey even when the mobile client application isn't running. In order to receive push notifications, users must first subscribe to them on the AppSettingsView page by turning on the push notifications **ToggleSwitch** before saving their application settings. In addition, users must then go to the FilterSettingsView page and select their desired tenants before saving their filter settings.

*Push notifications for Windows Phone can send notifications to users of your Windows Phone application even if the application isn't running. Toast notifications are ignored if the application is running, unless the **ShellToastNotification-Received** event is subscribed to. In this case the application can decide how it wants to respond to toast notifications.*

Figure 3 shows, at a high level, how this notification process works for the Tailspin Surveys application.

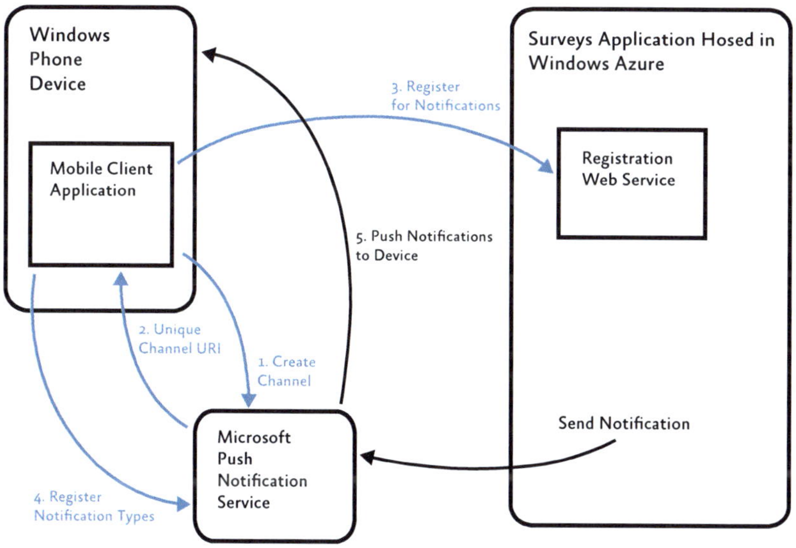

Figure 3
Push notifications for Windows Phone

Figure 3 shows how an application on the Windows Phone device can register for push notifications from another application—a service running in Windows Azure in this case. After the registration process is complete, the service in Windows Azure can push notifications to the Windows Phone device. The following describes the process illustrated in Figure 3:

1. The registration process starts when the client application establishes a channel by sending a message to the MPNS.

2. The MPNS returns a URI that is unique to the instance of the client application on a particular Windows Phone device.
3. Establishing the channel simply enables the phone to receive messages. The client application must also register with the service that will send the notification messages by sending its unique Uniform Resource Identifier (URI) to the service. In the Surveys application, there is a Registration web service hosted in Windows Azure that the mobile client application can use to register its URI for notifications of new surveys.
4. Provided that notification registration has succeeded, the mobile client application can also specify which types of notifications it will receive; this part of the registration process sets up a binding that enables the phone to associate a notification with the application and enables the user to launch the mobile client application in response to receiving a message. The Windows Phone Application Certification Requirements specify that you must provide the user with the ability to disable toast and tile notifications. You must run the application at least once to execute the code that establishes the channel before your phone can receive notifications.
5. Notifications are pushed to the Windows Phone device.

The service can use the unique URI to send messages to the client application. The Surveys service sends a message to a mobile client by sending a message to the endpoint specified by the URI that the client sent when it registered. The MPNS hosts this endpoint and forwards messages from the service on to the correct device.

For more information about the certification requirements that relate to push notifications, see Section 6.2, *"Push Notifications Application,"* of *"Additional Requirements for Specific Application Types"* on the MSDN® developer program website.

Remember that notifications are sent to the phone, not the application. This is because there is no guarantee that the application will be running when the Surveys service sends a message.

The sample application uses the free, unauthenticated MPNS that limits you to sending 500 notification requests per channel per day. If you use the free version of MPNS, it also means that your application is vulnerable to spoofing and denial of service attacks if someone impersonates the worker role that is sending notifications.

The authenticated MPNS has no restrictions on the number of notification messages you can send, and it requires the communication between your server component and MPNS to use Secure Sockets Layer (SSL) for sending notification messages.

For more information about the Microsoft Push Notification Service, see "Push Notifications Overview for Windows Phone" on MSDN.

Inside the Implementation

Now is a good time to take a more detailed look at the code that implements push notifications. As you go through this section, you may want to download the *Windows Phone Tailspin Surveys* application from the Microsoft Download Center.

Registering for Notifications

Before a phone can receive notifications from the Windows Azure service of new surveys, it must obtain its unique URI from the MPNS. This registration process takes place when the user taps the Save button on the AppSettingsView page.

The following code example shows the **IRegistrationServiceClient** interface in the TailSpin.PhoneClient project that defines the registration operations that the mobile client can perform.

C#
```
public interface IRegistrationServiceClient
{
  IObservable<TaskSummaryResult> UpdateReceiveNotifications(
    bool receiveNotifications);
  IObservable<Unit> UpdateTenants(
    IEnumerable<TenantItem> tenants);
  IObservable<SurveyFiltersInformation>
    GetSurveysFilterInformation();
  bool CredentialsAreInvalid();
  void CloseChannel();
  bool IsProcessing { get; }
  event EventHandler IsProcessingChanged;
}
```

For details of the **UpdateTenants** and **GetSurveysFilterInformation** methods, see the section, "Filtering Data," later in this chapter.

The **RegistrationServiceClient** class implements the **UpdateReceiveNotifications** method to handle the registration process for the mobile client. The following code example shows how the **UpdateReceiveNotifications** method handles the registration and unregistration processes in the mobile client application:

- If the user is enabling notifications from MPNS, the method creates an **HttpNotificationChannel** object and binds the channel to the toast and tile notifications, and registers the unique URI with the Tailspin Surveys service.

*The **HttpNotificationChannel** object stores the unique URI allocated by the MPNS.*

- If the user is disabling notifications from MPNS, the method unregisters from the Tailspin Surveys service and closes the channel.

C#
```
private const string ChannelName = "tailspindemo.cloudapp.net";
private const string ServiceName = "TailSpinRegistrationService";
...
public IObservable<TaskSummaryResult> UpdateReceiveNotifications(
  bool receiveNotifications)
{
  if (receiveNotifications)
  {
    httpChannel = new HttpNotificationChannel(ChannelName, ServiceName);
    ...
    // Bind the channel to the toast and tile nofiticaions and register
    // the URI with the Tailspin Surveys service.
    ...
  }
  else
  {
    httpChannel = HttpNotificationChannel.Find(ChannelName);

    if (httpChannel != null && httpChannel.ChannelUri != null)
    {
      return BindChannelAndUpdateDeviceUriInService(
        receiveNotifications, httpChannel.ChannelUri)
        .Select(taskSummary =>
        {
          IsProcessing = false;
          return TaskSummaryResult.Success;
        });
    }
    else
    {
      IsProcessing = false;
      return Observable.Return(TaskSummaryResult.Success);
    }
  }
}
```

When the user is enabling notifications, the code in the following example from the **UpdateReceiveNotifications** method creates the channel and registers the URI with the Tailspin Surveys service, and then binds the channel to the toast and tile notifications. It does this by creating the **HttpChannel** object and then converting the **ChannelUriUpdated** and **ErrorOccurred** events of the **HttpChannel** object into observable sequences using the **Observable.FromEvent** method, before it opens the channel. Once the observable sequence for the **ChannelUriUpdated** event has a value, the **BindChannelAndUpdateDeviceUriInService** method is called. The method also uses a timeout on the **ChannelUriUpdated** observable sequence, which throws a **TimeoutException** if the sequence doesn't get a value in 60 seconds. Finally, it returns the observable **TaskSummaryResult** object from whichever of the two observable sequences gets a value first.

C#

```
var channelUriUpdated =
  from evt in httpChannel.ObserveChannelUriUpdatedEvent()
  from result in BindChannelAndUpdateDeviceUriInService(receiveNotifications,
    evt.EventArgs.ChannelUri)
  select result;

var channelUriUpdateFail =
  from o in httpChannel.ObserveErrorOccurredEvent()
  select TaskSummaryResult.UnknownError;

httpChannel.Open();

// If the notification service does not respond in time, it is assumed that the
// server is unreachable. The first event that happens is returned.

return channelUriUpdated.Timeout(
  TimeSpan.FromSeconds(60)).Amb(channelUriUpdateFail).Take(1)
  .Select(tsr =>
   {
     IsProcessing = false;
     return tsr;
   })
  .Catch<TaskSummaryResult, TimeoutException>(
  (e) =>
  {
    IsProcessing = false;
    return Observable.Return(TaskSummaryResult.UnreachableServer);
  });
```

*The **Take(TSource)** method returns a specified number of contiguous values from the start of an observable sequence.*

The following code example shows how the **BindChannelAndUpdateDeviceUriInService** method in the **RegistrationServiceClient** class registers the clients unique URI with the Tailspin Surveys web service by asynchronously invoking a web method and passing it a **DeviceDto** object that contains the phone's unique URI. In addition, this method also binds the channel to the toast and tile notifications.

```
C#
private IObservable<TaskSummaryResult>
  BindChannelAndUpdateDeviceUriInService(
  bool receiveNotifications, Uri channelUri)
{
  var device = new DeviceDto
  {
    Uri = channelUri != null ? channelUri.ToString() : string.Empty,
    RecieveNotifications = receiveNotifications
  };

  var uri = new Uri(serviceUri, "Notifications");

  return httpClient
    .PostJson(new HttpWebRequestAdapter(uri), settingsStore.UserName,
    settingsStore.Password, device)
    .Select(u =>
    {
      BindChannelAndNotify(receiveNotifications);
      return TaskSummaryResult.Success;
    });
}
```

This method uses an instance of the **HttpClient** class to post the data transfer object to the web service. Tailspin developed the class to simplify the sending of asynchronous HTTP requests from the mobile client application.

The following code example shows how the **BindChannelAndNotify** method in the **Registration-ServiceClient** class configures the phone to respond to toast and tile notifications.

```
C#
private readonly Uri serviceUri;
private readonly IHttpClient httpClient;
...

private void BindChannelAndNotify(bool receiveNotifications)
{
  if (httpChannel != null)
  {
    if (receiveNotifications)
    {
      if (!httpChannel.IsShellToastBound)
```

```
        httpChannel.BindToShellToast();

      if (!httpChannel.IsShellTileBound)
        httpChannel.BindToShellTile();
    }
    else
    {
      if (httpChannel.IsShellToastBound)
        httpChannel.UnbindToShellToast();

      if (httpChannel.IsShellTileBound)
        httpChannel.UnbindToShellTile();
    }
  }
}
```

The following code example shows the **PostJson** method from the **HttpClient** class that uses the **Observable.FromAsyncPattern** method from the Reactive Extensions (Rx) framework to call the web method asynchronously. This code example shows four steps:

1. It first creates the **IHttpWebRequest** object and uses the **FromAsyncPattern** method to create an asynchronous function that returns an observable **Stream** object from the **IHttp-WebRequest** object.
2. It uses the **WriteContentToStream** method to attach the payload to the request stream.
3. It then calls the **BeginGetResponse** and **EndGetResponse** methods on the request object and returns an **IObservable<WebResponse>**.
4. The method returns an **IObservable<Unit>** instance, which is equivalent to a **null** in Rx, when it has a complete HTTP response message.

C#

```
public IObservable<Unit> PostJson<T>(IHttpWebRequest httpWebRequest,
  string userName, string password, T obj)
{
  var request = GetRequest(httpWebRequest, userName, password);
  request.Method = "POST";
  request.ContentType = "application/json";

  return from requestStream in Observable
    .FromAsyncPattern<Stream>(request.BeginGetRequestStream,
    request.EndGetRequestStream)()
    from response in WriteContentToStream(requestStream, request, obj)
    select new Unit();
}

private IObservable<WebResponse> WriteContentToStream<T>(Stream requestStream,
  IHttpWebRequest request, T obj)
{
```

```
  using (requestStream)
  {
    var serializer = new DataContractJsonSerializer(typeof(T));
    serializer.WriteObject(requestStream, obj);
  }

  return Observable.FromAsyncPattern<WebResponse>(
  request.BeginGetResponse, request.EndGetResponse)();
}
```

Figure 4 outlines the chain of observables that are initiated by the **AppSettingsViewModel-Submit** method.

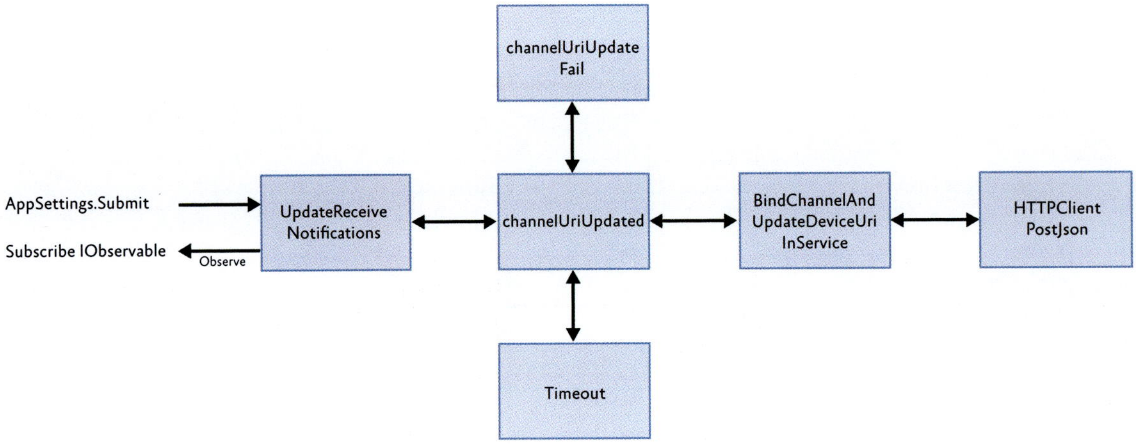

Figure 4
The chain of observables initiated by the AppSettingsViewModel Submit method.

The **Submit** method calls **UpdateReceiveNotifications** on the **IRegistrationServiceClient** instance, observes the result on the UI thread, and subscribes to the result. The call to **UpdateReceive-Notifications** results in one of four outcomes:

- **TaskSummaryResult.Success** is returned. In this case, the settings store is updated and if the intention was to turn off push notifications, then both the **IRegistrationServiceClient** instance and the subscription are disposed of.
- **TaskSummaryResult.UnreachableServer** is returned. This means that the update has timed out.
- **TaskSummaryResult.UnknownError** is returned. In this case the **ChannelUriUpdateFailed** event is raised by the **HttpNotificationChannel** instance.
- An exception is thrown. This may be a W**ebException** related to the HTTP post.

These outcomes are handled by the delegates passed into the call to the **Subscribe** method. This is the only call to the **Subscribe** method that is required to start the chain of observables. The **CloseChannel** method in the **RegistrationServiceClient** class closes and disposes of the **HttpChannel** object, and is called from the CleanUp method in the **AppSettingsViewModel** class. For more information see, "Handling Asynchronous Interactions" in Chapter 3, "Using Services on the Phone."

So far, this section has described Tailspin's implementation of the client portion of the registration process. The next part of this section describes how the Surveys service stores the unique URI that the client obtained from the MPNS in Windows Azure storage whenever a Windows Phone device registers for notifications of new surveys.

The following code example shows the implementation of the registration web service that runs in Windows Azure. You can find this class is in the Tailspin.Services.Surveys project.

C#

```
[AspNetCompatibilityRequirements(RequirementsMode =
  AspNetCompatibilityRequirementsMode.Allowed)]
[ServiceBehavior(InstanceContextMode =
  InstanceContextMode.PerCall)]
public class RegistrationService : IRegistrationService
{
  ...
  private readonly IUserDeviceStore userDeviceStore;
  ...
  public void Notifications(DeviceDto device)
  {
    var username = Thread.CurrentPrincipal.Identity.Name;

    bool isWellFormedUriString = Uri.IsWellFormedUriString(
      device.Uri, UriKind.Absolute);
    if (isWellFormedUriString)
    {
      if (device.RecieveNotifications)
      {
        this.userDeviceStore.SetUserDevice(username, new Uri(device.Uri));
      }
      else
      {
        this.userDeviceStore.RemoveUserDevice(new Uri(device.Uri));
      }
    }
  }
}
```

The **Notifications** method receives a **DeviceDto** parameter object that includes the unique URI allocated to the phone by the MPNS and a Boolean flag to specify whether the user is subscribing or unsubscribing to notifications. The service saves the details in the **DeviceDto** object in the device store in Windows Azure storage.

In the sample application, the service does not protect the data as it reads and writes to Windows Azure storage. When you deploy your application to Windows Azure, you should secure your table, binary large object (blob), and queue endpoints using SSL.

Although phones can explicitly unsubscribe from notifications, the application also removes devices from the device store if it detects that they are no longer registered with the MPNS when it sends a notification.

Sending Notifications

The Tailspin Surveys service sends toast notifications of new surveys to subscribed Windows Phone devices.

When a survey creator saves a new survey, the Surveys service in Windows Azure retrieves all the URIs for the subscribed Windows Phone devices and then sends the notifications to all the devices that subscribe to notifications for surveys created by that particular survey creator.

The following code example from the **NewSurveyNotificationCommand** class in the TailSpin.Workers.Notifications project shows how the Windows Azure worker role retrieves the list of subscribed phones and sends the notifications. This method uses the filtering service to retrieve the list of devices that should receive notifications about a particular survey. For more information, see the section, "Filtering Data," later in this chapter.

C#
```
public void Run(NewSurveyMessage message)
{
  var survey = this.surveyStore.GetSurveyByTenantAndSlugName(
    message.Tenant, message.SlugName, false);

  if (survey != null)
  {
    var deviceUris =
      from user in this.filteringService.GetUsersForSurvey(survey)
      from deviceUri in this.userDeviceStore.GetDevices(user)
      select deviceUri;

    foreach (var deviceUri in deviceUris)
    {
      this.pushNotification.PushToastNotification(
        deviceUri.ToString(),
        "New Survey", "tap 'sync' to get it",
         uri => this.userDeviceStore.RemoveUserDevice(
          new Uri(uri)));
    }
  }
}
```

The **Run** method also passes a reference to a callback method that removes from the device store phones that are no longer registered with the MPNS.

> *For more information about how the **Run** method is triggered, and about retry policies if the **Run** method throws an exception, see the section "The Worker Role 'Plumbing' Code" in Chapter 4, "Building a Scalable, Multi-Tenant Application for Windows Azure," of the book, Developing Applications for the Cloud on the Microsoft Windows Azure™ Platform 2nd Edition. This is available on MSDN.*

The following code example shows the **PushToastNotification** method in the **PushNotification** class, from the TailSpin.Web.Survey.Shared project, which is invoked by the worker role command to send the notification. This method creates the toast notification message to send to the MPNS before it calls the **SendMessage** method.

The **SendMessage** method acquires the stream and writes to it asynchronously because the service may be sending messages to hundreds or even thousands of devices, and for every device, it needs to open and write to a stream.

C#

```
public void PushToastNotification(string channelUri, string text1,
  string text2, DeviceNotFoundInMpns callback)
{
  byte[] payload = ToastNotificationPayloadBuilder.Create(
    text1, text2);
  string messageId = Guid.NewGuid().ToString();
  this.SendMessage(NotificationType.Toast, channelUri, messageId,
    payload, callback);
}
```

The **SendMessage** method sends messages to the MPNS for forwarding on to the subscribed devices. The following code example shows how the **SendMessage** method sends the message to the MPNS; the next code example shows how the **SendMessage** method receives a response from the MPNS.

C#

```
protected void OnNotified(NotificationType notificationType,
  HttpWebResponse response)
{
  var args = new NotificationArgs(notificationType, response);
  ...
}

private void SendMessage(NotificationType notificationType,
  string channelUri, string messageId, byte[] payload,
  DeviceNotFoundInMpns callback)
```

```
{
  try
  {
    WebRequest request = WebRequestFactory.CreatePhoneRequest(
      channelUri, payload.Length, notificationType, messageId);
    request.BeginGetRequestStream(
      ar =>
      {
        // Once async call returns get the Stream object
        Stream requestStream = request.EndGetRequestStream(ar);

        // and start to write the payload to the stream
        // asynchronously.
        requestStream.BeginWrite(
          payload,
          0,
          payload.Length,
          iar =>
          {
            // When the writing is done, close the stream
            requestStream.EndWrite(iar);
            requestStream.Close();

            // and switch to receiving the response from MPNS

            ...
          },
          null);
    },
    null);
  }
  catch (WebException ex)
  {
    if (ex.Status == WebExceptionStatus.ProtocolError)
    {
      this.OnNotified(notificationType, (HttpWebResponse)ex.Response);
    }
    Trace.TraceError(ex.TraceInformation());
  }
}
```

After the **SendMessage** method sends the message to the MPNS, it waits for a response. It must receive the message asynchronously because the MPNS does not return a response until it, in turn, receives a response from the Windows Phone device. The **SendMessage** method notifies its caller of the response through the **OnNotified** method call.

C#

```
...
// Switch to receiving the response from MPNS.
request.BeginGetResponse(
  iarr =>
  {
    try
    {
      using (WebResponse response = request.EndGetResponse(iarr))
      {
        // Notify the caller with the MPNS results.
        this.OnNotified(notificationType, (HttpWebResponse)response);
      }
    }
    catch (WebException ex)
    {
      if (ex.Status == WebExceptionStatus.ProtocolError)
      {
        this.OnNotified(notificationType, (HttpWebResponse)ex.Response);
      }
      if (((HttpWebResponse)ex.Response).StatusCode == HttpStatusCode.NotFound)
      {
          callback(channelUri);
      }
      Trace.TraceError(ex.TraceInformation());
    }
  },
  null);
...
```

If the **SendMessage** method receives a "404 Not Found" response code from the MPNS, it removes the stored subscription details from the store because this response indicates that the device is no longer registered with the MPNS.

The following table summarizes the information available from the MPNS in the response.

Item	Description
Message ID	This is a unique identifier for the response message.
Notification Status	This is the status of the notification message. Possible values are **Received**, **Dropped**, or **QueueFull**. You could use this value to determine whether you need to resend the message to this device.
Device Connection Status	This is the connection status of the device. Possible values are **Connected**, **InActive**, **Disconnected**, and **TempDisconnected**. If it is important that a device receive the message, you can use this value to determine whether you need to resend the message later.
Subscription Status	This is the device's subscription status. Possible values are **Active** and **Expired**. If a device's subscription has expired, you can remove its URI from your list of subscribed devices.

For more information about sending push notifications, see *"How to: Send a Push Notification for Windows Phone"* on MSDN.

Notification Payloads

The elements of a toast notification are:

- A title string that displays after the application icon.
- A content string that displays after the title.
- A parameter value that is not displayed but is passed to the application if the user taps on the toast notification; for example, the page the application should launch to, or name-value pairs to pass to the application.

For information about the elements of a tile notification you should read the section, "Using Live Tiles on the Phone," in Chapter 3, "Using Services on the Phone."

You must make sure that the strings you send on toast notifications fit in the available space.

The following code example from the **ToastNotificationPayloadBuilder** class in the TailSpin.Workers.Notifications project shows how the Surveys service constructs a toast notification message.

C#

```
public static byte[] Create(string text1, string text2 = null)
{
  using (var stream = new MemoryStream())
  {
    var settings = new XmlWriterSettings
    {
      Indent = true,
      Encoding = Encoding.UTF8
    };
    using (XmlWriter writer = XmlWriter.Create(stream, settings))
    {
      if (writer != null)
      {
        writer.WriteStartDocument();
        writer.WriteStartElement("wp", "Notification",
```

```
            "WPNotification");
          writer.WriteStartElement("wp", "Toast", "WPNotification");
          writer.WriteStartElement("wp", "Text1", "WPNotification");
          writer.WriteValue(text1);
          writer.WriteEndElement();
          writer.WriteStartElement("wp", "Text2", "WPNotification");
          writer.WriteValue(text2);
          writer.WriteEndElement();
          writer.WriteEndElement();
          writer.WriteEndDocument();
          writer.Close();
        }

        byte[] payload = stream.ToArray();
        return payload;
      }
    }
}
```

Accessing Data in the Cloud

The Surveys application stores its data in the cloud using a combination of Windows Azure tables and blobs. The mobile client application needs to access this data. For example, it must be able to retrieve survey definitions before it can display the questions to the phone user, and it must be able to save completed survey responses back to the cloud where they will be available for analysis by the survey creator.

Goals and Requirements

The mobile client application must be able to reliably download survey definitions and reliably upload survey responses. Making sure that surveys download reliably is important because survey creators want to be sure that surveys are delivered to all potential surveyors in order to maximize the number of responses. Making sure that surveys upload reliably is important because survey creators want to receive the maximum number of completed surveys, with no duplication of results.

The developers at Tailspin are aware that some surveyors may have limited bandwidth available, so they wanted to control the amount of bandwidth used to transfer data between the phone and the service. In addition to this, the developers at Tailspin want to make sure that the application performs well when it transfers data to and from the cloud.

The developers also wanted a solution that was as simple as possible to implement, and that they could easily customize in the future if, for example, authentication requirements were to change.

Finally, again with a view to the future, the developers wanted a solution that could potentially work with platforms other than Windows Phone.

Overview of the Solution

Tailspin considered three separate aspects of the solution: how to implement the server, how to implement the client, and the format of the data that the application moves between the phone and the cloud.

Exposing the Data in the Cloud

Tailspin is using the WCF REST programming model to exchange data between the mobile client and the cloud-based service.

The developers at Tailspin decided to use the Windows Communication Foundation (WCF) Representational State Transfer (REST) programming model to expose the data in the Tailspin Surveys service.

Data Formats

Because Tailspin is using a custom WCF service, they must use custom data transfer objects to exchange data between the mobile client application and the cloud-based service. Tailspin chose to use a JavaScript Object Notation (JSON) format for moving the data over the wire because it produces a compact payload that reduces bandwidth requirements, is relatively easy to use, and will be usable on platforms other than the Windows Phone platform.

> *If, in the future, Tailspin moves to WCF Data Services, it will no longer require the custom data transfer objects because WCF Data Services use the OData protocol to move data over the wire.*

Tailspin also considered compressing the data before transferring it over the network to reduce bandwidth utilization, but the developers at Tailspin decided that the additional CPU and battery usage on the phone that this would require outweighed the benefits in this particular case. You should evaluate this tradeoff between the cost of bandwidth and battery consumption in your own application before you decide whether to compress data you need to move over the network.

Consuming the Data

The developers at Tailspin implemented a set of custom classes in the mobile client application to handle the data transfer with the Tailspin web service. The classes on the mobile client interact with the WCF REST service and parse the data received from the service. Tailspin's analysis of the data transfer requirements for the Windows Phone application identified only two types of interaction with the service: a "Get Surveys" operation and a "Send Survey Result" operation, so the implementation of a custom client should be quite simple. Furthermore, the "Send Survey Result" operation always appends the result to the store on the server so there are no concurrency issues, and survey creators cannot modify a survey design after they publish it so there are no versioning issues.

In the future, Tailspin may decide to use WCF Data Services and the OData protocol. OData client libraries, including a version for Windows Phone, are available for download on the ***Open Data Protocol*** *website.*

Using the OData Client Library would minimize the amount of code that the developers would have to write because the library and the code generated by using the DataSvcUtil utility fully encapsulate the calls to the WCF Data Service endpoint. Furthermore, using the client library offers advanced features such as batching, client-side state management, and conflict resolution.

For a walkthrough that shows how to use the OData Client Library on the Windows Phone platform, see the post, ***"Walkthrough: Consuming OData with MVVM for Windows Phone,"*** *on MSDN.*

Tailspin also evaluated the Microsoft Sync Framework to handle the synchronization of data between the phone and Windows Azure, but it was decided that the simplicity of the synchronization required by the Tailspin mobile client application did not warrant this.

Using SSL

Self-signed certificates are not supported on the Windows Phone device, so to implement SSL, it is necessary to use a server certificate from a trusted third-party company, such as VeriSign. Therefore, the sample application does not secure the WCF REST service with SSL, so a malicious client can impersonate the phone client and send malicious data.

You should protect any sensitive data that you need to transfer over the network between the Windows Phone client and the Windows Azure-hosted services by using SSL.

Inside the Implementation

Now is a good time to take a more detailed look at the code that enables the mobile client application to access data in the cloud. As you go through this section, you may want to download the *Windows Phone Tailspin Surveys* application from the Microsoft Download Center.

Creating a WCF REST Service in the Cloud

The TailSpin.Services.Surveys project includes a standard WCF REST Service named SurveysService hosted in Windows Azure that exposes the survey data to the Windows Phone client application. The Windows Azure web role defined in the TailSpin.Services.Surveys.Host.Azure project populates a routing table that includes a route to the Surveys service in its Global.asax.cs file. The following code example shows how the **RegisterRoutes** method creates the **RouteTable** object.

```
C#
public class Global : HttpApplication
{
  protected void Application_Start(object sender, EventArgs e)
  {
    ...
    RegisterRoutes();
  }

  private static void RegisterRoutes()
  {
    var customServiceHostFactory = new
      CustomServiceHostFactory(ContainerLocator.Container);
    RouteTable.Routes.Add(new ServiceRoute("Registration",
      customServiceHostFactory, typeof(RegistrationService)));
    RouteTable.Routes.Add(new ServiceRoute("Survey",
      customServiceHostFactory, typeof(SurveysService)));
  }
}
```

This **SurveysService** class implements the WCF REST service endpoints. The following code example shows the **GetSurveys** method in the **SurveysService** class that exposes the surveys data stored in Windows Azure storage.

```
C#
public SurveyDto[] GetSurveys(string lastSyncUtcDate)
{
  DateTime fromDate;
  if (!string.IsNullOrEmpty(lastSyncUtcDate))
  {
    if (DateTime.TryParse(lastSyncUtcDate, out fromDate))
    {
      fromDate = DateTime.SpecifyKind(fromDate,
        DateTimeKind.Utc);
    }
    else
    {
      throw new FormatException("lastSyncUtcDate is in an
        incorrect format. The format should be:
          yyyy-MM-ddTHH:mm:ss");
    }
  }
  else
  {
```

```
    fromDate = new DateTime(1900, 1, 1, 0, 0, 0, DateTimeKind.Utc);
  }

  var username = Thread.CurrentPrincipal.Identity.Name;

  return this.filteringService
    .GetSurveysForUser(username, fromDate)
    .Select(s => new SurveyDto
    {
      SlugName = s.SlugName,
      Title = s.Title,
      Tenant = s.Tenant,
      Length = 5 * s.Questions.Count,
      IconUrl = this.GetIconUrlForTenant(s.Tenant),
      CreatedOn = s.CreatedOn,
      Questions = s.Questions.Select(q => new QuestionDto
      {
        PossibleAnswers = q.PossibleAnswers,
        Text = q.Text,
        Type = Enum.GetName(typeof(QuestionType), q.Type)
      }).ToList()
    }).ToArray();
}
```

This example shows how the Surveys service returns survey definitions to the mobile client in an array of **SurveyDto** objects that represent surveys added to the service after a specified date. It also demonstrates how you can apply a filter to the request. In the current version of the application, the filter returns a list of surveys from the phone user's preferred list of tenants. For more information about how Tailspin implemented the filtering behavior, see the section, "Filtering Data," later in this chapter.

To enable the mobile client to upload survey answers, the **SurveysService** class provides two methods: one for uploading individual images and sound clips, and one for uploading complete survey answers. The following code example shows how the **AddMediaAnswer** method saves an image or a sound clip to Windows Azure blob storage and returns a URI that points to the blob.

C#

```
public string AddMediaAnswer(Stream media, string type)
{
  var questionType = (QuestionType)Enum.Parse(typeof(QuestionType), type);
  return this.mediaAnswerStore.SaveMediaAnswer(media, questionType);
}
```

The following code example shows the **AddSurveyAnswers** method that receives an array of **Survey-AnswerDto** objects that it unpacks and saves in the survey answer store in Windows Azure storage.

```
C#
public void AddSurveyAnswers(SurveyAnswerDto[] surveyAnswers)
{
  foreach (var surveyAnswerDto in surveyAnswers)
  {
    this.surveyAnswerStore.SaveSurveyAnswer(new SurveyAnswer
    {
      Title = surveyAnswerDto.Title,
      SlugName = surveyAnswerDto.SlugName,
      Tenant = surveyAnswerDto.Tenant,
      StartLocation = surveyAnswerDto.StartLocation,
      CompleteLocation = surveyAnswerDto.CompleteLocation,
      QuestionAnswers = surveyAnswerDto.QuestionAnswers
      .Select(qa => new QuestionAnswer
      {
        QuestionText = qa.QuestionText,
        PossibleAnswers = qa.PossibleAnswers,
        QuestionType = (QuestionType)Enum.Parse(
          typeof(QuestionType), qa.QuestionType),
        Answer = qa.Answer
      }).ToList()
    });
  }
}
```

Consuming the Data in the Windows Phone Client Application

The mobile client application uses the methods exposed by the Surveys service to send and receive survey data.

The following code example shows the **ISurveysServiceClient** interface that defines the set of asynchronous WCF REST calls that the mobile client application can make.

```
C#
public interface ISurveysServiceClient
{
  IObservable<IEnumerable<SurveyTemplate>> GetNewSurveys(string lastSyncDate);
  IObservable<Unit> SaveSurveyAnswers(IEnumerable<SurveyAnswer> surveyAnswers);
}
```

The **SurveysServiceClient** class implements this interface, and the following code example shows the **GetNewSurveys** method that sends the request to the service and returns the observable list of surveys to the application. This method makes the asynchronous web request by using the **GetJson** method from the **HttpClient** class, converts the returned data transfer objects to **Survey-Template** objects, and then returns an observable sequence of **SurveyTemplate** objects.

```
C#
public IObservable<IEnumerable<SurveyTemplate>>
  GetNewSurveys(string lastSyncDate)
{
  var surveysPath = string.Format(CultureInfo.InvariantCulture,
    "Surveys?lastSyncUtcDate={0}", lastSyncDate);
  var uri = new Uri(this.serviceUri, surveysPath);

  return
    httpClient
      .GetJson<IEnumerable<SurveyDto>>(
        new HttpWebRequestAdapter(uri),
        settingsStore.UserName, settingsStore.Password)
      .Select(ToSurveyTemplate);
}
```

*For more information about the **HttpClient** class that makes the asynchronous web request, see the section, "Registering for Notifications," earlier in this chapter.*

To save completed survey answers to the Surveys service, the client must first save any image and sound-clip answers. It uploads each media answer in a separate request, and then it includes the URL that points to the blob that contains the media when the client uploads the complete set of answers for a survey. It would be possible to upload the survey answers and the media in a single request, but this may require a large request that exceeds the maximum upload size configured in the service. By uploading all the media items first, the worst that can happen, if there is a failure, is that there are orphaned media items in Windows Azure storage.

The following code example shows the first part of the **SaveAndUpdateMediaAnswers** method that creates a list of answers that contain media.

```
C#
var mediaAnswers =
  from surveyAnswer in surveyAnswersDto
  from answer in surveyAnswer.QuestionAnswers
  where answer.Answer != null &&
    (answer.QuestionType ==
    Enum.GetName(typeof(QuestionType), QuestionType.Picture) ||
    answer.QuestionType ==
    Enum.GetName(typeof(QuestionType), QuestionType.Voice))
  select answer;
```

Tailspin decided to support clients that upload individual media items instead of using multi-part messages in order to support the widest set of possible client platforms.

The method then iterates over this list and asynchronously creates HTTP requests to post each media item to the Tailspin Surveys service.

```
C#
foreach (var answer in mediaAnswers)
{
  var mediaAnswerPath = string.Format(CultureInfo.InvariantCulture,
    "MediaAnswer?type={0}", answer.QuestionType);
  var mediaAnswerUri = new Uri(this.serviceUri, mediaAnswerPath);
  byte[] mediaFile = GetFile(answer.Answer);

  var request = httpClient.GetRequest(
    new HttpWebRequestAdapter(mediaAnswerUri),
    settingsStore.UserName, settingsStore.Password);
  request.Method = "POST";
  request.Accept = "application/json";
  ...
}
```

The method then asynchronously sends each of these requests and retrieves the response to each request in order to extract the URL that points to the blob where the service saved the media item. It must then assign the returned URLs to the **Answer** property of the correct **QuestionAnswerDto** data transfer object. The following code example shows how the **SaveAndUpdateMediaAnswers** method sends the media answers to the Tailspin Surveys service asynchronously. The code example shows how this operation is broken down into the following steps:

1. The method first makes an asynchronous call to access the HTTP request stream as an observable object.
2. The method writes the media item to the request stream and then makes an asynchronous call to access the HTTP response stream.
3. It reads the URL of the saved media item from the response stream and sets the **Answer** property of the **QuestionAnswerDto** object.
4. The **ForkJoin** method makes sure that the media answers are uploaded in parallel, and after all the media answers are uploaded, the method returns an **IObservable<Unit>** object.

```
C#
var mediaAnswerObservables = new List<IObservable<Unit>>();
foreach (var answer in mediaAnswers)
{
  ...

  QuestionAnswerDto answerCopy = answer;
  var saveFileAndUpdateAnswerObservable = Observable
    .FromAsyncPattern<Stream>(request.BeginGetRequestStream,
    request.EndGetRequestStream)()
    .SelectMany(requestStream =>
    {
```

```
      using (requestStream)
      {
        requestStream.Write(mediaFile, 0, mediaFile.Length);
        requestStream.Close();
      }
      return Observable.FromAsyncPattern<WebResponse>(
        request.BeginGetResponse, request.EndGetResponse)();
    },
    (requestStream, webResponse) =>
    {
      using (var responseStream = webResponse.GetResponseStream())
      {
        var responseSerializer = new
          DataContractJsonSerializer(typeof(string));
        answerCopy.Answer =
         (string)responseSerializer.ReadObject(responseStream);
      }

      return new Unit();
    });

  mediaAnswerObservables.Add(saveFileAndUpdateAnswerObservable);
}

return mediaAnswerObservables.ForkJoin().Select(u => new Unit());
```

The following code example shows how the mobile client uploads completed survey answers. It first creates the data transfer objects, then it uploads any media answers using the **SaveAndUpdateMediaAnswers** method described earlier, and finally, it uploads the **SurveyAnswerDto** object using the **HttpClient** class. The **SelectMany** method here makes sure that all the media answers are uploaded before the **SurveyAnswerDto** object.

C#

```
public IObservable<Unit>
  SaveSurveyAnswers(IEnumerable<SurveyAnswer> surveyAnswers)
{
  var surveyAnswersDto = ToSurveyAnswersDto(surveyAnswers);

  var saveAndUpdateMediaAnswersObservable =
    SaveAndUpdateMediaAnswers(surveyAnswersDto);

  var uri = new Uri(this.serviceUri, "SurveyAnswers");
  var saveSurveyAnswerObservable =
    httpClient.PostJson(new HttpWebRequestAdapter(uri),
    settingsStore.UserName, settingsStore.Password,
    surveyAnswersDto);
```

```
    return saveAndUpdateMediaAnswersObservable.SelectMany(
        u => saveSurveyAnswerObservable);
}
```

Both the **GetNewSurveys** and **SaveSurveyAnswer** methods are called from the **SurveysSynchronizationService** class that is responsible for coordinating which surveys should be downloaded and which survey answers should be uploaded. For a description of the synchronization process, see Chapter 3, "Using Services on the Phone."

Filtering Data

The Surveys service filters lists of surveys for notifications and for synchronization.

Users of the mobile client application can express preferences for which surveys they would like to see on their devices. In the application, the criteria will be a list of preferred tenants (survey creators such as Adatum and Fabrikam in the sample). There are two scenarios in which the Tailspin Surveys service must filter data based on filter criteria from the client:

- The cloud service can send notifications to mobile clients to tell them when new surveys are available. The notification service should only send a notification to a user if one of the user's preferred tenants created the new survey.
- When the mobile client synchronizes its data with the Tailspin Surveys service, the phone should only receive surveys created by the user's preferred tenants.

For more details about the notification process, see the section, "Notifying the Mobile Client of New Surveys," earlier in this chapter. For more details about the synchronization process, see the section, "Synchronizing Data between the Client and the Cloud," in Chapter 3, "Using Services on the Phone."

Overview of the Solution

There are three tasks that make up the filtering functionality in the Tailspin Surveys application: registering the user's preferences; identifying which devices to notify when a tenant adds a new survey; and identifying which surveys to include in the synchronization process.

Registering User Preferences

The user creates a list of preferred tenants on the FilterSettingsView page in the mobile client application. When the user saves the settings, the Tailspin Surveys mobile client application sends the preferences to the Tailspin Surveys registration service that then stores the list of preferred tenants in Windows Azure table storage.

Figure 5 outlines the way that the mobile client application saves the user's settings in the Tailspin Surveys web service.

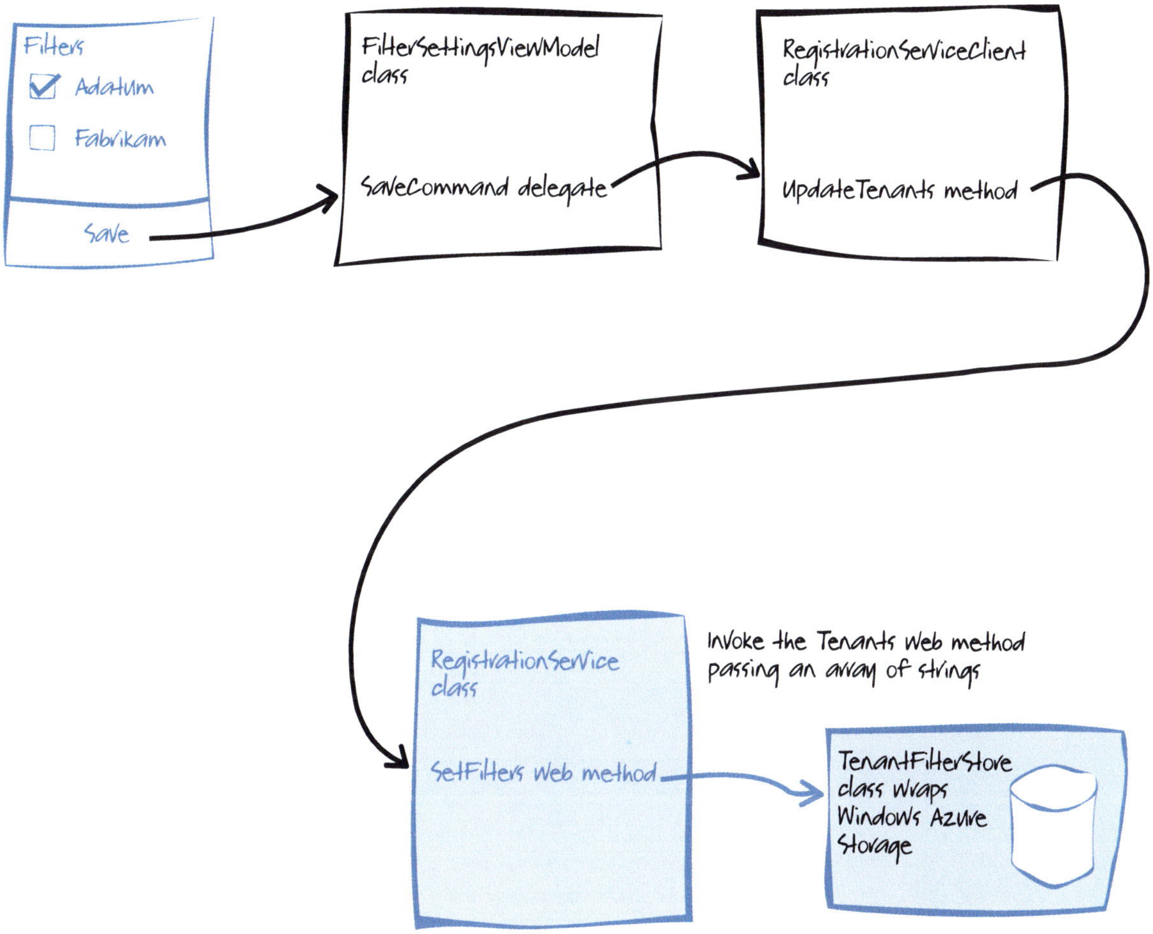

Figure 5
Saving user settings in the Tailspin Surveys web service

The view model uses a service class to send the settings to the Tailspin Surveys registration web service. All the settings are encapsulated in a data transfer object that the web service code unpacks and saves in Windows Azure table storage.

The following table shows the structure of the tenant filter table in Windows Azure table storage that stores the list of preferred tenants for each user.

Column	Notes
Tenant name	A string holding the tenant name. This column is the table's partition key.
User name	A string holding the user name. This column is the table's row key.

The application can use this table to retrieve a list of tenants that a particular user is interested in or a list of users who are interested in a particular tenant.

*For more information about Windows Azure table storage, including partition keys and row keys, see Chapter 5, "**Phase 2: Automating Deployment and Using Windows Azure Storage,**" of the book, **Moving Applications to the Cloud 2nd Edition**. It is available on MSDN.*

The following table shows the structure of the user device table in Windows Azure table storage that records which physical devices, identified by a unique URI allocated by the MPNS, are associated with each user.

Column	Notes
User name	A string holding the user name. This column is the table's partition key.
Device URI	A string holding the device's unique URI that the MPNS uses to push notifications to the device. This column is the table's row key.

The structure of this table reflects the fact that a user can have more than one device, each with a unique URI for the MPNS to use. Tailspin optimized this table to support queries that look for the list of devices owned by a user.

With the current implementation, it's possible for a user who owns multiple Windows Phone devices to enter different lists of preferred tenants on the different devices. The data model in the Tailspin Surveys service only supports a single list of preferred tenants for each user, so the user might see unexpected results on some of their devices.

Identifying Which Devices to Notify

Figure 6 outlines the process that the Tailspin Surveys service uses to identify which devices to notify when a subscriber adds a new survey.

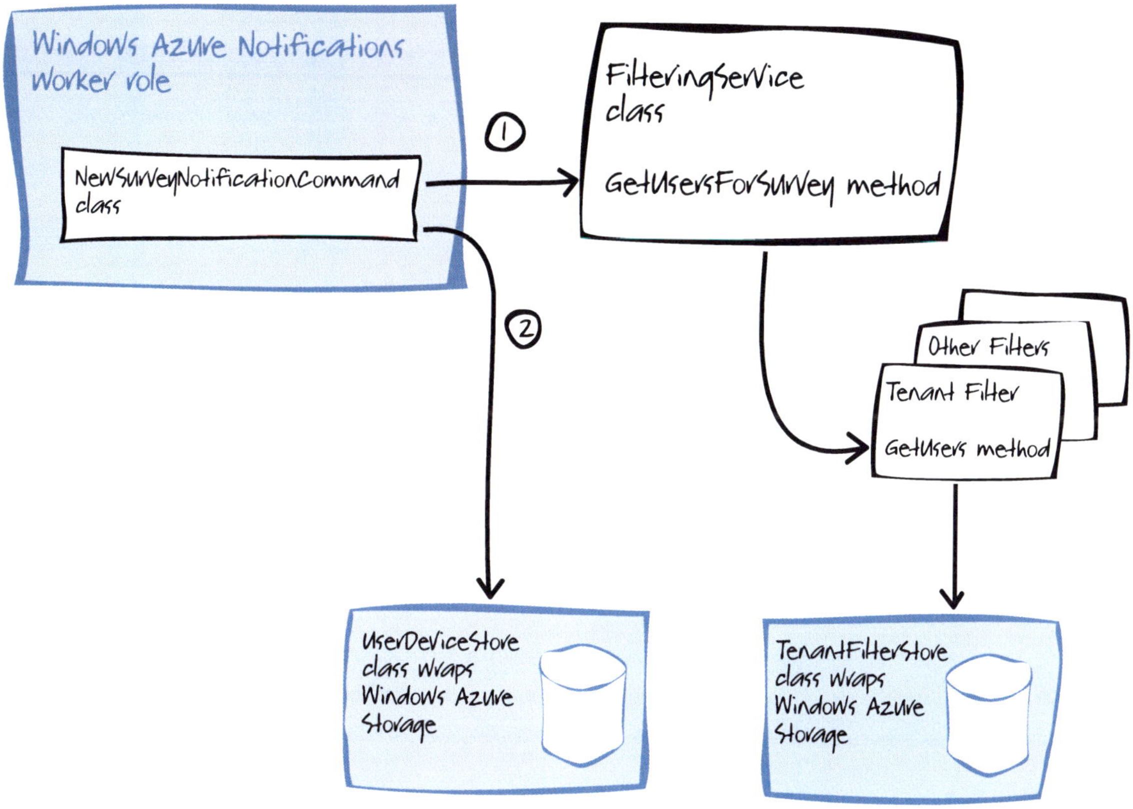

Figure 6
Building a list of devices to notify about a new survey

Identifying which devices to notify when a tenant adds a new survey requires two queries. The first query gets the list of users who are interested in that particular survey and this query uses a filtering service to build this list of users based on user preferences stored in Windows Azure table storage.

The second query uses the list of users to find a list of devices to notify, again using data stored in Windows Azure table storage that maps users to devices.

In the current version of the application, obtaining the list of users who are interested in a particular survey uses just a single criterion: the user's list of preferred tenants. In the future, the criteria may become more complex if, for example, users can express preferences for location, survey length, or some other attribute of the survey. The developers at Tailspin have implemented an extensible filtering mechanism to accommodate any future additions to the list of supported criteria.

Selecting Surveys to Synchronize

Figure 7 outlines the way that the Tailspin application delivers new, relevant surveys to the mobile client application when it synchronizes with the Tailspin Surveys service. The user initiates the synchronization process from the SurveyListView page, and the view model invokes a method in a service class to retrieve the list of new surveys. Alternatively, the synchronization process can be automatically initiated by the background agent, which periodically runs if the user has enabled background tasks in the AppSettingsView page. In the Tailspin Surveys service, the application gets the list of surveys to return to the client from a filtering service class.

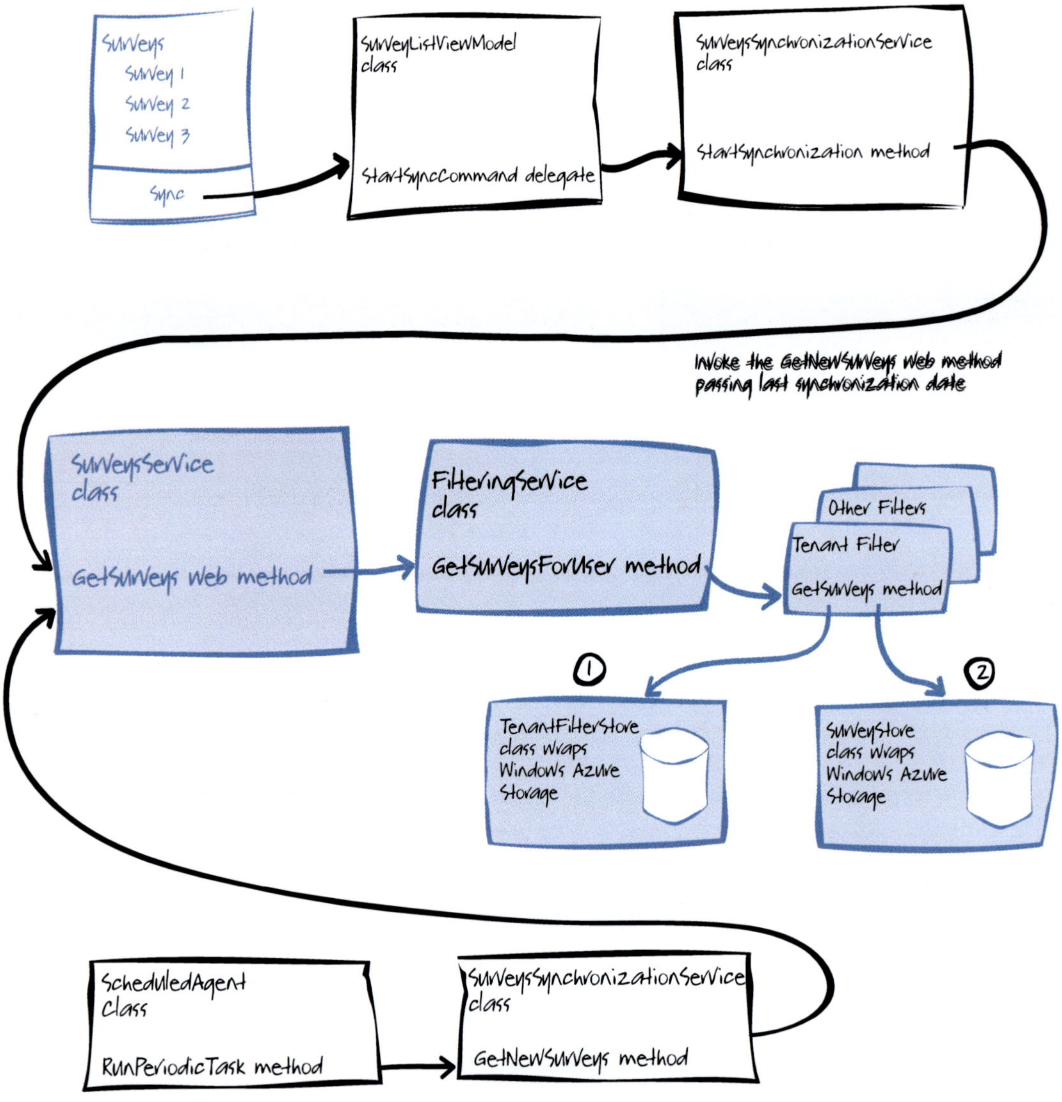

FIGURE 7
Building a list of new surveys for a user

Identifying which surveys to download when a user synchronizes from the mobile client requires two queries. The first query returns a list of tenants based on user preferences stored in Windows Azure table storage. In the current version of the application, these preferences are lists of tenants in which the user is interested. The second query, which is inside the filter, uses the list of tenants to return a list of new surveys from those tenants.

Inside the Implementation

Now let's take a more detailed tour of the code that implements the filtering of the list of surveys. As you go through this section, you may want to download the *Windows Phone Tailspin Surveys* application from the Microsoft Download Center.

Storing Filter Data

The Tailspin Surveys service stores filter data in Windows Azure table storage. It uses one table to store the tenant filter data, and one table to store the device filter data. The following code example shows how the **TenantFilterStore** class saves tenant filter data in the Windows Azure table storage.

C#

```
public class TenantFilterStore : ITenantFilterStore
{
  private readonly IAzureTable<TenantFilterRow>
    tenantFilterTable;
  ...
  public void SetTenants(string username,
    IEnumerable<string> tenants)
  {
    var rowsToDelete = (from r in this.tenantFilterTable.Query
      where r.RowKey == username
      select r).ToList();

    this.tenantFilterTable.Delete(rowsToDelete);

    var rowsToAdd = tenants.Select(t => new TenantFilterRow
    {
      PartitionKey = t,
      RowKey = username
    });
    this.tenantFilterTable.Add(rowsToAdd);
  }
}
```

The following code example shows how the **UserDeviceStore** class saves device data.

```
C#
public class UserDeviceStore : IUserDeviceStore
{
  private readonly IAzureTable<UserDeviceRow> userDeviceTable;
  ...
  public void SetUserDevice(string username, Uri deviceUri)
  {
    this.RemoveUserDevice(deviceUri);

    var encodedUri = Convert.ToBase64String(
      Encoding.UTF8.GetBytes(deviceUri.ToString()));
    this.userDeviceTable.Add(new UserDeviceRow
    {
      PartitionKey = username,
      RowKey = encodedUri
    });
  }

  public void RemoveUserDevice(Uri deviceUri)
  {
    var encodedUri = Convert.ToBase64String(
      Encoding.UTF8.GetBytes(deviceUri.ToString()));
    var row = (from r in this.userDeviceTable.Query
      where r.RowKey == encodedUri
      select r).SingleOrDefault();

    if (row != null)
    {
      this.userDeviceTable.Delete(row);
    }
  }
}
```

The following code example shows how the **SetFilters** web method in the **RegistrationService** class stores the list of preferred tenants sent from the mobile client.

```
C#
public void SetFilters(SurveyFiltersDto surveyFiltersDto)
{
  var username = Thread.CurrentPrincipal.Identity.Name;

  this.tenantFilterStore.SetTenants(username,
    surveyFiltersDto.Tenants.Select(t => t.Key));
}
```

*There is also a **GetFilters** web method to return the current list of preferred tenants to the client. The phone does not save the filter information locally; instead, it retrieves it from the web service using the **GetFilters** method when it needs it.*

Building a List of Devices to Receive Notifications

Whenever a tenant creates a new survey, the Tailspin Surveys service must send notifications to all the devices that are interested in that survey. The criteria that the service uses to select the devices may change in future versions of the application, so the developers at Tailspin implemented an extensible filtering method to build the list.

The following code example from the Run method in the **NewSurveyNotificationCommand** class retrieves a list of device URIs to which it will send notifications. The **NewSurveyNotificationCommand** class implements a command that a Windows Azure worker role executes in response to receiving a message on a Windows Azure queue.

C#

```
var deviceUris =
  from user in this.filteringService.GetUsersForSurvey(survey)
  from deviceUri in this.userDeviceStore.GetDevices(user)
  select deviceUri;
```

This code uses the **FilteringService** class to get a list of users who are interested in the new survey, and then it uses the **GetDevices** method in the **UserDeviceStore** class to get a list of device URIs for those users. The application populates the user device store when the mobile client registers for notifications with the MPNS. For more information, see the section, "Notifying the Mobile Client of New Surveys," earlier in this chapter.

The following code example shows how the **FilteringService** class uses one or more **ISurveyFilter** instances in the **GetUsersForSurvey** method to filter the list of users.

C#

```
private readonly ISurveyFilter[] filters;
...
public IEnumerable<string> GetUsersForSurvey(Survey survey)
{
  return (from filter in this.filters
    from user in filter.GetUsers(survey)
    select user).Distinct();
}
```

The current version of the application uses a single filter class named **TenantFilter** to retrieve a list of users from the tenant filter store based on the tenant name. The following code example shows part of this **TenantFilter** class.

```
C#
public class TenantFilter : ISurveyFilter
{
  ...
  private readonly ITenantFilterStore tenantFilterStore;
  ...
  public IEnumerable<string> GetUsers(Survey survey)
  {
    return this.tenantFilterStore.GetUsers(survey.Tenant);
  }
}
```

Building a List of Surveys to Synchronize with the Mobile Client

Whenever the synchronization process is initiated, the Tailspin Surveys service must build a list of new surveys that the user is interested in. The current version of the application uses the user's list of preferred tenants to select the list of surveys, but Tailspin may expand the set of selection criteria in the future. The mobile client invokes the **GetSurveys** web method in the **SurveysService** class to get a list of new surveys. The **GetSurveys** method uses the **GetSurveysForUser** method of the **FilteringService** class to get the list of surveys that it will return to the client.

The following code example shows how the **FilteringService** class gets the list of surveys for a user.

```
C#
private readonly ISurveyFilter[] filters;
...
public IEnumerable<Survey> GetSurveysForUser(
  string username, DateTime fromDate)
{
  return (from filter in this.filters
    from survey in filter.GetSurveys(username, fromDate)
    select survey).Distinct(new SurveysComparer());
}
```

This method can use one or more **ISurveyFilter** objects to filter the list of surveys. The current version of the application uses a single filter named **TenantFilter**. The following code example shows the **GetSurveys** method of the **TenantFilter** class that first retrieves a list of tenants for the user, and then retrieves a list of surveys for those tenants.

```
C#
public IEnumerable<Survey> GetSurveys(string username, DateTime fromDate)
{
  var tenants = this.tenantFilterStore.GetTenants(username);
  if (tenants.Count() == 0)
  {
    tenants = this.tenantStore.GetTenants().Select(t =>
      t.Name.ToLowerInvariant());
  }

  return (from tenant in tenants
    from survey in this.surveyStore.GetSurveys(tenant, fromDate)
    select survey).Distinct(new SurveysComparer());
}
```

Conclusion

This chapter described the ways in which the Tailspin Surveys mobile client application accesses remote services; it also discussed designing services that Windows Phone can access. You have now seen a complete picture of the Tailspin Surveys mobile client application, including the application's UI, Tailspin's use of the MVVM pattern, how the application leverages services offered by the Windows Phone platform, and how it consumes services over the network.

Questions

1. How does Tailspin pass authentication requests to the web service?
    a. Tailspin uses basic authentication with the credentials in an authorization header.
    b. Tailspin uses Window Live ID.
    c. Tailspin uses OAuth.
    d. Tailspin uses the Windows Identity Framework (WIF).
2. What notification methods does the Microsoft Push Notification Service support?
    a. Toast notifications.
    b. Tile notifications.
    c. SMS notifications.
    d. Raw notifications.
3. Which of the following are elements of a toast notification?
    a. A title string that displays after the application icon.
    b. A content string that displays after the title.
    c. A background image.
    d. A parameter value that is not displayed but is passed to the application if the user taps on the toast.

4. Why does the client need to register with MPNS before it can receive notifications?
   a. Because MPNS requires clients to authenticate before it will send notifications.
   b. Because MPNS can then notify your service that the client is ready to receive notifications.
   c. Because the client must obtain a unique URI to send to your service.
   d. Because the free version of MPNS has a limit on the number of clients that can receive notifications from your service.
5, How does Tailspin transport data between the client and the web service?
   a. Tailspin uses the Microsoft Sync Framework to handle the data transport.
   b. Tailspin uses the WCF Data Services framework.
   c. Tailspin uses data transfer objects with a WCF REST endpoint.
   d. The mobile client application uploads directly to Windows Azure BLOB storage.
6. Why does Tailspin filter data on the server and not on the client?
   a. To minimize the amount of data moved over the network.
   b. To simplify the application.
   c. For security reasons.
   d. To minimize storage requirements on the phone.

More Information

For more information about push notifications, see *"Push Notifications for Windows Phone"* on MSDN.

For more information about developing websites for Windows Phone, see *"Web Development for Windows Phone"* on MSDN.

For more information about security for Windows Phone, see *"Security for Windows Phone"* on MSDN.

These and all links in this book are accessible from the book's online bibliography. You can find the bibliography on MSDN at: *http://msdn.microsoft.com/en-us/library/gg490786.aspx*.

APPENDIX A Unit Testing Windows Phone Applications

Unit tests differ from integration or user acceptance tests in that they test the smallest unit of functionality possible. Typically, a unit test will test the behavior of a specific method. The table below compares unit and integration tests.

Unit Tests	**Integration Tests**
Dependent systems are not required	Dependent systems are required and must be online
Fast to run	Slow to run
Typically run frequently (during development and code check-in)	Typically run periodically
Tests developer expectations	Tests real system behavior
No fear of refactoring as the functionality should stay the same and the unit tests should continue to pass	No fear of refactoring as the functionality should stay the same and the unit tests should continue to pass

The most important aspect of unit testing compared to integration testing is that unit tests should not be reliant on the functionality of dependent systems to exercise the code being tested. If the method being tested handles a condition such as an **IsolatedStorageException** being thrown, it would not be ideal to fill up or somehow corrupt isolated storage in order to trigger the code path. Similarly, testing code that calls services or queries databases should not be reliant on dependent systems. If a test requires dependent systems to be correctly configured, running, and populated with a known set of data, the test would be very fragile and would fail if any of these conditions were not met. In such circumstances you would see many failing tests because of issues with dependent systems, rather than issues with the code under test.

A good approach to increase software testability is to isolate dependent systems and have them passed into your business logic using an abstraction such as an interface or abstract class. This approach allows the dependent system to be passed into the business logic at run time. In addition, in the interests of testability, it also allows a mock version of the dependent system to be passed in at test time. For more information see, *"Forms of Dependency Injection,"* on Martin Fowler's website.

As previously mentioned, unit tests should run quickly. The Tailspin mobile client application unit tests run in less than 10 minutes and are triggered on every check-in to the source control system. Since the unit tests run quickly, Tailspin developers know very quickly whether their check-in caused problems. Integration tests typically take longer to run and are usually scheduled less frequently. In addition to helping catch problems with code integration, unit tests are perhaps most valuable when refactoring code. Unit tests can be thought of as the code-based documentation of how the code under test should behave. If the code under test is refactored or reorganized, the functionality should stay the same and the unit tests should continue to pass. If the unit tests don't pass, it is possibly due to improperly refactored code. In addition, another possibility is that the desired code behavior has changed, in which case the unit tests should be updated or new tests should be added.

Windows Phone 7.1 SDK Abstractions

There are few public interfaces or abstract classes available in the Windows® Phone 7.1 SDK. For this reason, the Tailspin developers created their own. An interface was generated for each class in the Windows Phone 7.1 SDK that was used in the Tailspin mobile client application. Then, adapters and facades, which are wrapper classes that pass parameters and return values to and from the underlying Windows Phone 7.1 SDK classes, were created.

Figure 1 shows how the **CameraCaptureTask** class in the **Microsoft.Phone.Tasks** namespace is adapted by the **CameraCaptureTaskAdapter** class in the **TailSpin.PhoneClient.Adapters** namespace.

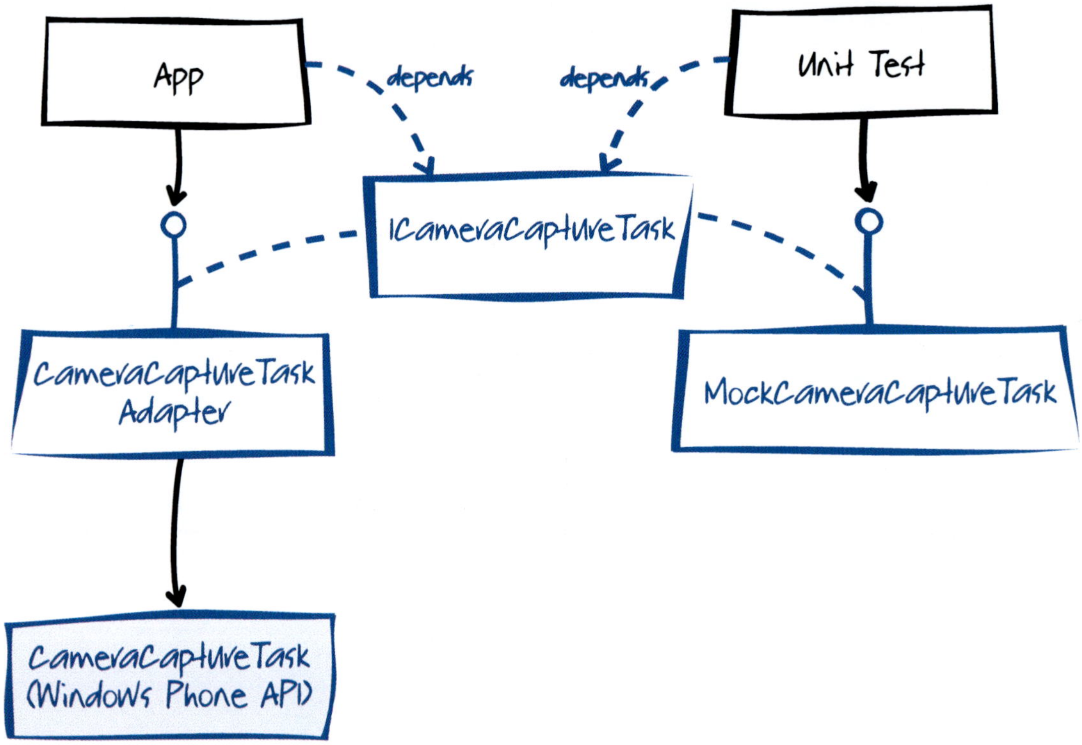

FIGURE 1
Adapting the CameraCaptureTask class

The **CameraCaptureTaskAdapter** class implements the **ICameraCaptureTask** interface from the **TailSpin.PhoneClient.Adapters** namespace. At run time, the Tailspin mobile client application uses the **CameraCaptureTaskAdapter** class directly, rather than the **CameraCaptureTask** class.

The right hand side of Figure 1 shows the **MockCameraCaptureTask** class from the **TailSpin.Phone.TestSupport** namespace. The **MockCameraCaptureTask** class also implements the **ICamera-CaptureTask** interface and is used in unit tests instead of the **CameraCaptureTaskAdapter** class. This allows Tailspin developers to unit test business logic that needs to interact with the camera.

The interfaces and adapters that abstract the Windows Phone 7.1 SDK classes can be found in the Tailspin.Phone.Adapters and Tailspin.PhoneClient.Adapters projects.

Mock Implementations

Unit tests should focus on how the code under test functions in response to values returned by dependent systems. By using mocks, the return values or exceptions to be thrown by mock instances of dependent systems can easily be controlled. For more information see, *"Exploring the Continuum of Test Doubles,"* in *MSDN Magazine*.

The TailSpin.Phone.TestSupport project contains mock implementations of the Windows Phone 7.1 SDK adapter classes created by Tailspin. The mock classes were manually developed as it is not possible to use a mocking framework on the Windows Phone platform. Mocking frameworks require the ability to emit Microsoft intermediate language (MSIL) code, which is not currently possible on the Windows Phone platform. The mocks were developed to be general-purpose implementations with many of them having properties that accept delegates. Delegate-accepting properties enable the execution of any desired behavior necessary for the unit test. The following code example shows the **MockProtectDataAdapter** class.

C#

```
public class MockProtectDataAdapter : IProtectData
{
  public Func<byte[], byte[], byte[]> ProtectTestCallback { get; set; }

  public Func<byte[], byte[], byte[]> UnProtectTestCallback { get; set; }

  public byte[] Protect(byte[] userData, byte[] optionalEntropy)
  {
    if(ProtectTestCallback == null)
    {
      throw new InvalidOperationException("Must set ProtectTestCallback.");
    }
    return ProtectTestCallback(userData, optionalEntropy);
  }

  public byte[] Unprotect(byte[] encryptedData, byte[] optionalEntropy)
  {
    if (UnProtectTestCallback == null)
    {
      throw new InvalidOperationException("Must set UnProtectTestCallback.");
```

```
    }
    return UnProtectTestCallback(encryptedData, optionalEntropy);
  }
}
```

The **MockProtectDataAdapter** class is an example of how a general purpose mock can be given behavior using delegates. The **ProtectTestCallback** property should be set with a delegate or a lambda expression so that a call to the **Protect** method can execute the delegate. By initializing the mock in this way, unit tests have unlimited test control of the mock.

Testing Asynchronous Functionality

The following code example shows the **ViewModelGetsPictureFromCameraTask** test method, which demonstrates testing asynchronous functionality. The unit test validates that the **Picture-QuestionViewModel** class can get an image from the camera capture task and set the view model's **Picture** property to the returned image. The **Asynchronous** method attribute allows the test to run until an unhandled exception is thrown, or until the **EnqueueTestComplete** method is called.

C#

```
[TestMethod, Asynchronous]
public void ViewModelGetsPictureFromCameraTask()
{
  var mockCameraCaptureTask = new MockCameraCaptureTask();
  WriteableBitmap picture = null;
  var imageUri = new Uri("/TailSpin.PhoneClient.Tests;
    component/ViewModels/Images/block.jpg", UriKind.Relative);
  var src = new BitmapImage();
  src.SetSource(Application.GetResourceStream(imageUri).Stream);
  picture = new WriteableBitmap(src);

  var sri = Application.GetResourceStream(imageUri);
  mockCameraCaptureTask.TaskEventArgs = new SettablePhotoResult
  {
    ChosenPhoto = sri.Stream
  };

  var questionAnswer = new QuestionAnswer
  {
    QuestionText = "Will this test pass?"
  };

  var target = new PictureQuestionViewModel(questionAnswer,
    mockCameraCaptureTask, new MockMessageBox());
  Assert.IsNull(target.Picture);

  target.PropertyChanged += (s, e) =>
  {
```

```
    if (e.PropertyName != "Picture") return;

    Assert.IsNotNull(target.Picture);
    Assert.IsTrue(IntArraysMatch(target.Picture.Pixels, picture.Pixels));

    EnqueueTestComplete();
  };

  target.CameraCaptureCommand.Execute();
}
```

The test first configures an instance of the **MockCameraCaptureTask** class, setting its **TaskEventArgs** property with a **SettablePhotoResult** that contains a specific picture. The test listens to the view model's **PropertyChanged** event looking for a change to the **Picture** property. In the event handler for the **PropertyChanged** event, the picture is validated by comparing the pixels of the source and target images.

The **Execute** method call on the view model's **CameraCaptureCommand** triggers the code under test. The view model calls the **Show** method on the **MockCameraCaptureTask** class and then listens for the **Completed** event. Once the event occurs and a picture is returned, the **Picture** property in the view model is set.

The unit test validates this behavior but must wait for the **MockCameraCaptureTask Completed** event to be fired, and the **PictureQuestionViewModel PropertyChanged** event to be fired, in order to validate the results.

Using Delegates to Specify Behavior

The following code example shows the **SettingPasswordUsesEncryptionServiceAndPersistsIntoIsoStore** test method, which demonstrates how to initialize a mock by passing a delegate to it. The unit test validates that the SettingsStore class encrypts passwords prior to isolated storage persistence.

C#

```
[TestMethod]
public void SettingPasswordUsesEncryptionServiceAndPersistsIntoIsoStore()
{
  var encoder = new UTF8Encoding();
  var mockProtectDataAdapter = new MockProtectDataAdapter();
  mockProtectDataAdapter.ProtectTestCallback =
    (userData, optionalEntropy) =>
    {
      var stringUserData = encoder.GetString(userData, 0, userData.Length);
      return encoder.GetBytes(string.Format("ENCRYPTED: {0}",
        stringUserData));
    };
  var target = new SettingsStore(mockProtectDataAdapter);
  IsolatedStorageSettings.ApplicationSettings["PasswordSetting"] = null;
```

```
    target.Password = "testpassword";

    var encryptedByteArray = (byte[])IsolatedStorageSettings
      .ApplicationSettings["PasswordSetting"];
    Assert.AreEqual("ENCRYPTED: testpassword",
      encoder.GetString(encryptedByteArray, 0, encryptedByteArray.Length));
}
```

The test first configures an instance of the **MockProtectDataAdapter** class, providing it behavior that will execute when its **Protect** method is called. In this case, the behavior is simply to prepend the string "ENCRYPTED:" to the **Password** property. The test triggers the code by setting the **Password** property on the **SettingsStore** class. The test then validates that the value that persists into isolated store was encrypted by the behavior defined in the test.

Running Unit Tests

Tailspin runs unit tests on the emulator and on a real device to make sure the test behavior is not affected by any behavioral differences in the core libraries on the phone as compared to the desktop. It would be even more valuable if these unit tests could be automatically run every time the code base is altered. In order to automate the running of unit tests, the developers at Tailspin use a custom Microsoft Build Engine (MSBuild) task and a custom logger.

This MSBuildTasks.RunWP7UnitTestsInEmulator task performs the following steps:

1. Connects to the Windows Phone emulator
2. Installs the unit test application
3. Launches the unit test application
4. Loads the test results file from the isolated storage in the emulator
5. Examines the test results file for failing tests

In order to capture the test results into a file, a custom logger is used. This FileLogProvider class in the TailSpin.Phone.TestSupport project extends Microsoft.Silverlight.Testing.Harness.LogProvider and captures all messages into a text file which is saved to isolated storage.

All links in this book are accessible from the book's online bibliography. You can find the bibliography on MSDN at: ***http://msdn.microsoft.com/en-us/library/gg490786.aspx***.

APPENDIX B

Prism Library for Windows Phone

Prism is a free library from the Microsoft patterns & practices group. The components in this library can help developers build applications for Windows® Presentation Foundation (WPF), Microsoft Silverlight® browser plug-in, and the Windows Phone platform that are easier to maintain and update as requirements change.

Prism was originally built to support composite application scenarios, where you might typically have a shell application and modules that contribute pieces to the overall application. Although Prism supports this application style, most of the library components offered by Prism can be useful in building any WPF or Silverlight application. For example, Prism offers components to support the Model-View-ViewModel (MVVM) pattern as well as pieces to provide loosely coupled communication between parts of an application.

Although the use of a display shell is typically not appropriate in a Windows Phone application, many of the other components within Prism are useful when building Silverlight applications for Windows Phone. For example, you can use the **DelegateCommand** provided by Prism to avoid the requirement to implement event handlers in the code-behind of your views when using the MVVM pattern for your applications.

Prism includes a small library known as the Prism Library for Windows Phone, which contains a subset of the main Prism Library features specifically aimed at helping developers implement solutions to common issues found in developing applications for Windows Phone. The library includes classes to help developers implement commands, navigation, observable object notifications, data template selection, interaction with notifications, interaction with the application bar, and more.

This appendix provides an overview of the Prism Library for Windows Phone, which is used in the Tailspin application for Windows Phone discussed in this guide. For more information and to download Prism, see the *Prism* home page on the MSDN® developer program website. To provide feedback, get assistance, or download additional content, visit the *Prism* community site on CodePlex.

About Prism for Windows Phone

Windows Phone applications implemented using Silverlight are naturally suited to the MVVM pattern, which discourages the use of code-behind in the views in favor of handling events and activity in the view model. However, many common scenarios—such as binding commands to interface objects, linking methods to application bar buttons, notifying of changes to object properties, and detecting changes to text-based controls in the view—are challenging to accomplish without using code-behind. The helper classes and components in Prism Library for Windows Phone are specially designed to simplify these tasks in Silverlight applications created for Windows Phone.

In addition, the Prism Library for Windows Phone includes helper classes for publishing and subscribing to events, displaying notification messages, and selecting data templates at run time. Many of these components and helper classes are used in the Tailspin.PhoneClient project. You can download the complete *Windows Phone Tailspin Surveys* application for use in conjunction with this guide from the Microsoft Download Center.

The Prism Library for Windows Phone is provided as source code in the two projects, Microsoft.Practices.Prism and Microsoft.Practices.Prism.Interactivity, within the Tailspin solution. These solutions implement the Prism Library for Windows Phone library, and there is also a set of tests to help you explore the use of the classes in the library. These tests can also be used if you extend or modify the source code to meet your own specific requirements.

Contents of Prism for Windows Phone Library

The Prism Library for Windows Phone contains several namespaces that subdivide the artifacts:

- **Microsoft.Practices.Prism**. This namespace contains classes concerned with detecting and reacting to change events for properties and objects.
- **Microsoft.Practices.Prism.Commands**. This namespace contains classes concerned with binding commands to user interface (UI) objects without requiring the use of code-behind in the view, and composing multiple commands.
- **Microsoft.Practices.Prism.Events**. This namespace contains classes concerned with subscribing to events and with publishing events on the publisher thread, UI thread, or a background thread.
- **Microsoft.Practices.Prism.ViewModel**. This namespace contains classes that help support implementation of the view model portion of the MVVM pattern, such as displaying a template at run time and simplifying the implementation of property change notification.
- **Microsoft.Practices.Prism.Interactivity**. This namespace contains classes and custom behaviors concerned with handling interaction and navigation for application bar buttons and with updating the values of view model properties bound to text and password controls.

- **Microsoft.Practices.Prism.Interactivity.InteractionRequest**. This namespace contains classes concerned with displaying notifications to users.

The following tables list the main classes in these namespaces and provide a brief description of their usage. They do not include all the classes in each namespace; they list only those that implement the primary functions of the library. For a full reference of all the Prism namespaces, see *"Prism"* on MSDN.

Microsoft.Practices.Prism Namespace

The following table lists the main components in the **Microsoft.Practices.Prism** namespace. These classes are not used in the Tailspin application.

Class	Description
ObservableObject<T>	A class that wraps an object so that other classes can be notified of change events. Typically, this class is set as a dependency property on dependency objects and allows other classes to observe any changes in the property values. This class is required because, in Silverlight, it is not possible to receive change notifications for dependency properties that you do not own.
ExtensionMethods (for **List<T>**)	This class adds the **RemoveAll** method to **List<T>**. The method removes the all the elements that match the conditions defined by the specified predicate.

For more information about using the **ObservableObject** class, see the section *"Sharing Data Between Multiple Regions,"* in Chapter 7, "Composing the User Interface" in *"Prism"* on MSDN.

Microsoft.Practices.Prism.Commands Namespace

The following table lists the main components in the **Microsoft.Practices.Prism.Commands** namespace.

Class	Description	Usage in Tailspin application
CompositeCommand	This class composes one or more **ICommand** implementation instances. It forwards to all the registered commands, and returns **true** if all the commands return **true**.	This class is not used in the Tailspin application.
DelegateCommand	This class is an implementation of **ICommand**. Its delegates can be attached to access the **Execute** and **CanExecute** methods.	Used in most of the view models for associating commands with actions that execute in response to a command being fired. For more information, see the section, "Commands," in Chapter 2, "Building the Mobile Client."

For more information about using the **DelegateCommand** and **CompositeCommand** classes, see the section, *"Commands"* in Chapter 5, "Implementing the MVVM Pattern," in *"Prism"* on MSDN.

Microsoft.Practices.Prism.Events Namespace

The following table lists the main components in the **Microsoft.Practices.Prism.Events** namespace. Classes in this namespace support decoupled communication between pieces of an application, such as between two view models. These classes are not used in the Tailspin application.

Class	Description
CompositePresentationEvent	A class that manages publication and subscription to events.
SubscriptionToken	The subscription token returned from an event subscription method. This class provides methods to compare tokens.
EventSubscription	This class provides a method for retrieving a **Delegate** that executes an Action depending on the value of a second filter predicate that returns true if the action should execute.
BackgroundEventSubscription	This class extends **EventSubscription** to invoke the **Action** delegate on a background thread.
DispatcherEventSubscription	This class extends **EventSubscription** to invoke the **Action** delegate in a specific **Dispatcher** instance.
EventAggregator	This class implements a service that stores event references and exposes the **GetEvent** method to retrieve a specific event type.

For more information about publishing and subscribing to events, see Chapter 9, *"Communicating between Loosely Coupled Components,"* in *"Prism"* on MSDN.

Microsoft.Practices.Prism.ViewModel Namespace

The following table lists the main components in the **Microsoft.Practices.Prism.ViewModel** namespace.

Class	Description	Usage in Tailspin application
DataTemplateSelector	This class implements a custom **Content-Control** that changes its **ContentTemplate** based on the content it is presenting. To determine the template it must use for the new content, the control retrieves it from its resources using the name for the type of the new content as the key.	Used in the TakeSurveyView view to select the appropriate question view, depending on the type of question. For more information, see the section, "Data Binding and the Pivot Control," in Chapter 2, "Building the Mobile Client."
NotificationObject	This is a base class for items that support property notification. It provides basic support for implementing the **INotify-PropertyChanged** interface and for marshaling execution to the UI thread.	Used in the **ViewModel**, **QuestionViewModel**, **QuestionOption**, and **QuestionAnswer** classes. For more information, see the section, "Displaying Data," in Chapter 2, "Building the Mobile Client."

For more information about using the **DataTemplateSelector** class, see *"MVVM QuickStart"* in *"Prism"* on MSDN.

Microsoft.Practices.Prism.Interactivity Namespace

The following table lists the main components in the **Microsoft.Practices.Prism.Interactivity** namespace.

Class	Description	Usage in Tailspin application
ApplicationBarButtonCommand	This class implements a behavior that associates a command with an application bar button.	Used in the SurveyListView, TakeSurveyView, AppSettingsView, and FilterSettingsView views to bind the click events for application bar buttons to methods in the view models. For more information, see the section, "Commands," in Chapter 2, "Building the Mobile Client."
ApplicationBarButtonNavigation	This class implements a behavior that subscribes to an application bar button click event and navigates to a specified page.	This class is not used in the Tailspin application.
ApplicationBarExtensions	This class provides the **FindButton** method to find a button on the phone's application bar.	The **FindButton** method is used by the **ApplicationBarButtonCommand** class to find a button by name.
UpdatePasswordBindingOnPropertyChanged	This class implements a behavior that updates the source of a binding on a password box as the text changes. By default in Silverlight, the bound property is only updated when the control loses focus.	Used in the AppSettingsView view to ensure that the user's password input is updated in the view model even if the password box does not lose focus. For more information, see the section, "Displaying Data," in Chapter 2, "Building the Mobile Client."
UpdateTextBindingOnPropertyChanged	This class implements a behavior that updates the source of a binding on a text box as the text changes. By default in Silverlight, the bound property is only updated when the control loses focus.	Used in the AppSettingsView and OpenQuestionView views to ensure that the user's text input is updated in the view model even if the text box does not lose focus. For more information, see the section, "Displaying Data," in Chapter 2, "Building the Mobile Client."

Microsoft.Practices.Prism.Interactivity.InteractionRequest Namespace

The following table lists the main components in the **Microsoft.Practices.Prism.Interactivity.InteractionRequest** namespace.

Class	Description	Usage in Tailspin application
InteractionRequest	This class represents a request for user interaction. View models can expose interaction request objects through properties and raise them when user interaction is required so that views associated with the view models can materialize the user interaction using an appropriate mechanism.	Used in the SurveyListViewModel, AppSettingsViewModel, and FilterSettingsViewModel view models to generate interaction requests that display notification messages. For more information, see the section, "User Interface Notifications," in Chapter 2, "Building the Mobile Client."
MessageBoxAction	This class displays a message box as a result of an interaction request.	Used in the SurveyListView, AppSettingsView, and FilterSettingsView views to display message boxes. For more information, see the section, "User Interface Notifications," in Chapter 2, "Building the Mobile Client."
ToastPopupAction	This class displays a pop-up toast item with specified content as a result of an interaction request. After a short period, it removes the pop-up window.	Used in the SurveyListView view to display information about the most recent synchronization with the remote service. For more information, see the section, "User Interface Notifications," in Chapter 2, "Building the Mobile Client."

For more information about interaction requests and displaying notifications, see Chapter 6, *"Advanced MVVM Scenarios"* in *"Prism"* on MSDN.

These and all links in this book are accessible from the book's online bibliography. You can find the bibliography on MSDN at: ***http://msdn.microsoft.com/en-us/library/gg490786.aspx***.

Answers to Questions

Chapter 2, Building the Mobile Client

1. Which of the following are good reasons to use the MVVM pattern for your Windows® Phone application?

    a. It improves the testability of your application.

    b. It facilitates porting of the application to another platform, such as the desktop.

    c. It helps to make it possible for designers and developers to work in parallel.

    d. It may help you avoid risky changes to existing model classes.

    Answer: (a) True. *This is a valid reason for using the MVVM pattern.* **(b) False**. *The view and view model are often tightly coupled and use features of the Windows Phone platform, making these elements difficult to port.* **(c) True**. *Designers can work on the XAML files that make up the view independently of the developers working on the code in the view models.* **(d) True**. *Sometimes you may have existing model objects that encapsulate complex domain logic that you don't want to change. In this case, the view model acts as an adaptor, exposing the model to the view.*

For more information, see the section, "Benefits of MVVM," in Chapter 2, "Building the Mobile Client."

2. Which of the following are good reasons not to use the MVVM pattern for your Windows Phone application?

    a. You have a very tight deadline to release the application.

    b. Your application is relatively simple with only two screens and no complex logic to implement.

    c. Windows Phone controls are not ideally suited to the MVVM pattern.

    d. It's unlikely that your application will be used for more than six months before it is completely replaced.

    Answer: (a) True. *The MVVM pattern tends to add complexity to the application, especially during the initial development. However, you should carefully evaluate the long-term benefits of using the MVVM pattern.* **(b) True**. *Such a simple application may not warrant* *the overhead of implementing the MVVM pattern.* **(c) False**. *The Windows Phone controls support binding in the same way as standard Microsoft® Silverlight® browser plugin controls. In some cases though, you may have to develop custom behaviors to get specific functionality.* **(d) True**. *One of the advantages of the MVVM pattern is improved maintainability. For an application with a short shelf life, in may not be worth the overhead of the development process associated with the MVVM pattern.*

For more information, see the section, "Benefits of MVVM," in Chapter 2, "Building the Mobile Client."

3. Which of the following are correct about tombstoning?
   a. Tombstoned applications have been terminated.
   b. Tombstoned applications remain intact in memory.
   c. Information about a tombstoned application's navigation state and state dictionaries are preserved for when the application is relaunched.
   d. A device will maintain tombstoning information for up to five applications at once.

   Answer: (a) True. *The tombstoning process does terminate an application.* **(b) False**. *The tombstoning process terminates an application; therefore it does not remain in memory.* **(c) True**. *When an application is tombstoned, information about its navigation state and state dictionaries populated by the application during the Deactivated event are preserved.* **(d) True**. *A Windows Phone device will maintain tombstoning information for up to five applications at a time.*

For more information, see the section, "The Structure of the Tailspin Surveys Client Application," in Chapter 2, "Building the Mobile Client."

4. Which of the following describe the role of the view model locator?
   a. The view model locator configures bindings in the MVVM pattern.
   b. In the Tailspin mobile client, the view model locator is responsible for instantiating view model objects.
   c. The view model locator connects views to view models.
   d. Data template relations offer an alternative approach to a view model locator.

   Answer: (a) False. *Although the view model locator establishes the link between a view and a view model, the bindings are configured in the view's XAML code.* **(b) False**. *In the Tailspin application, the dependency injection container instantiates the view model objects and manages their lifetimes.* **(c) True**. *This is the core functionality of the view model locator.* **(d) True**. *This is an alternative way to connect views to view models.*

For more information, see the section, "Connecting the View and the View Model," in Chapter 2, "Building the Mobile Client."

5. Where does the Back button take you?

   a. To the previous view in the navigation stack.
   b. It depends on what the code in the view model does.
   c. If the current view is the last one in the navigation stack, you leave the application.
   d. If your application is on the top of the phone's application stack, it takes you back to your application.

   Answer: (a) True. *This is what happens within your application.* **(b) False**. *The behavior of the hardware Back button is determined by the operating system.* **(c) True**. *This returns you to the phone's Start screen.* **(d) True**. *You can use the Back button to navigate back to your application from the phone's environment.*

For more information, see the section, "Page Navigation," in Chapter 2, "Building the Mobile Client."

6. Why should you not use code-behind when you're using the MVVM pattern?

   a. The view model locator always intercepts the events, so code-behind code never executes.
   b. The MVVM pattern enforces a separation of responsibilities between the view and the view model. UI logic belongs in the view model.
   c. If you are using the MVVM pattern, other developers will expect to see your code in the view model classes and not in the code-behind.
   d. Code-behind has a negative effect on view performance.

   Answer: (a) False. *The view model locator does not intercept control events.* **(b) True**. *This is the separation of responsibilities between the view and the view model.* **(c) True**. *This is the expected place for UI logic.* **(d) False**. *Whether code is in code-behind files or in view model classes has no effect on the application's performance.*

For more information, see the section, "Using the Model-View-View Model Pattern," in Chapter 2, "Building the Mobile Client."

Chapter 3, Using Services on the Phone

1. The Data Protection API (DPAPI) can be used to encrypt and decrypt data in isolated storage. What does the DPAPI use as an encryption key?

   a. A user-generated private key.
   b. The user credentials.
   c. The phone credentials.
   d. The user and phone credentials.

   Answer: (a) False. *The Windows Phone platform does not support generated private keys for encrypting and decrypting data in isolated storage.* **(b) False**. *User credentials alone are not used to encrypt and decrypt data in isolated storage.* **(c) False**. *Phone credentials alone* *are not used to encrypt and decrypt data in isolated storage.* **(d) True**. *The DPAPI solves the problem of explicitly generating and storing a cryptographic key by using the user and phone credentials to encrypt and decrypt data in isolated storage.*

For more information, see the section, "Security," in Chapter 3, "Using Services on the Phone."

2. What happens when your application is reactivated?

   a. You return to the first screen in your application.
   b. The operating system makes sure that the screen is displayed as it was when the application was deactivated.
   c. The operating system recreates the navigation stack within your application.
   d. The Launching event is raised.

   Answer: (a) False. *You are returned to the last screen that was visible before the application was deactivated.* **(b) False**. *It's your application's responsibility to restore whatever state is required to reset the application's appearance and behavior.* **(c) True**. *If you were several pages deep within the application when it was deactivated, you will still be several pages deep when the application is reactivated.* **(d) False**. *The Launching event is raised when the application is run from scratch, not when it is reactivated. There is an Activated event that is raised when the application is reactivated.*

For more information, see the section, "Handling Activation and Deactivation," in Chapter 3, "Using Services on the Phone."

3. What data should you save when you handle the deactivation request?

   a. State data required to rebuild the state of the last screen that was active before the application was deactivated.
   b. State data required to rebuild the state of previous screens that the user had navigated through before the application was deactivated.
   c. Data that is normally persisted to isolated storage by the application at some point.
   d. The currently active screen.

   Answer: (a) True. *It's the application's responsibility to manage the state data used to redisplay the UI, although the operating system will remember which page in your application was visible when it is deactivated.* **(b) True**. *It's the application's responsibility to manage the state data used to redisplay the UI, including any screens in the application's navigation stack.* **(c) True**. *There is no guarantee that an application will be reactivated, so you should save anything important.* **(d) False**. *The operating system will record which screen was active for you.*

For more information, see the section, "Handling Activation and Deactivation," in Chapter 3 "Using Services on the Phone."

4. Why does Tailspin use the Reactive Extensions (Rx) for .NET?
   a. To handle notifications from the Microsoft Push Notification Service.
   b. To handle UI events.
   c. To manage asynchronous tasks.
   d. To make the code that implements the asynchronous and parallel operations more compact and easier to understand.

   Answer: (a) False. *You don't need to use the Rx for this task.* **(b) False**. *You don't need to use the Rx for this task.* **(c) True**. *This is a core use of the Rx.* **(d) True**. *This is a benefit of using Rx.*

For more information, see the section, "Synchronizing Data between the Phone and the Cloud," in Chapter 3, "Using Services on the Phone."

5. What factors should you take into account when you use location services on the phone?
   a. The level of accuracy your application requires for its geo-location data.
   b. Whether the device has a built-in GPS.
   c. How quickly you need to obtain the current location.
   d. Whether the user has consented to your application using the phone's GPS data.

   Answer: (a) True. *You can specify the required level of accuracy, and this will affect how the phone obtains its current position: using GPS, or from triangulation.* **(b) False**. *The hardware specification for the Windows Phone device includes a GPS module.* **(c) True**. *Using GPS is often slower, though more accurate, than triangulation.* **(d) True**. *The guidelines for the phone state that you must obtain the user's consent before using geo-location data from the phone in your application.*

For more information, see the section, "Using Location Services on the Phone," in Chapter 3, "Using Services on the Phone."

6. Which factors constrain the use of a **ResourceIntensiveTask** agent?
   a. Resource-intensive agents do not run unless the Windows Phone device is connected to an external power source.
   b. Resource-intensive agents do not run unless the Windows Phone device has a network connection over Wi-Fi or through a connection to a PC.
   c. Resource-intensive agents do not run unless the Windows Phone device's battery power is greater than 90%.
   d. Resource-intensive agents do not run unless the Windows Phone device screen is locked.

   Answer: (a) True. *A Windows Phone device must be connected to an external power source for a ResourceIntensiveTask agent to run.* **(b) True**. *A Windows Phone device must have a network connection over Wi-Fi or through a connection to a PC for a ResourceIntensiveTask agent to run.* **(c) True**. *A Windows Phone device must have battery power of greater than 90% for a ResourceIntensiveTask agent to run.* **(d) True**. *A Windows Phone device screen must be locked for a ResourceIntensiveTask agent to run.*

For more information, see the section, "Synchronizing Data between the Phone and the Cloud," in Chapter 3, "Using Services on the Phone."

Chapter 4, Connecting with Services

1. How does Tailspin pass authentication requests to the web service?
   a. Tailspin uses basic authentication with the credentials in an authorization header.
   b. Tailspin uses Windows Live® ID.
   c. Tailspin uses OAuth.
   d. Tailspin uses the Windows Identity Framework (WIF).

   Answer: (a) True. *This is Tailspin's current approach.* **(b) False**. *However, this might be a mechanism for Tailspin to adopt in the future.* **(c) False**. *However, this might be a mechanism for Tailspin to adopt in the future.* **(d) False**. *Tailspin uses WIF to process the authentication request in the web service.*

For more information, see the section, "Authenticating with the Surveys Service," in Chapter 4, "Connecting with Services."

2. What notification methods does the Microsoft Push Notification Service (MPNS) support?
   a. Toast notifications.
   b. Tile notifications.
   c. SMS notifications.
   d. Raw notifications.

   Answer: (a) True. *Used for important notifications for immediate viewing, such as breaking news.* **(b) True**. *Used for informational notifications such as a temperature change for a weather application.* **(c) False**. *This is not a feature of MPNS.* **(d) True**. *You can use raw notifications in addition to tile and toast notifications to send information directly to your application.*

For more information, see the section, "Notifying the Mobile Client of New Surveys," in Chapter 4, "Connecting with Services."

3. Which of the following are elements of a toast notification?
   a. A title string that displays after the application icon.
   b. A content string that displays after the title.
   c. A background image.
   d. A parameter value that is not displayed but is passed to the application if the user taps on the toast.

   Answer: (a) True. *Toast notifications contain a boldface title string that displays immediately after the application icon.* **(b) True**. *Toast notifications include a non-boldface content* *string that displays immediately after the title.* **(c) False**. *This is not an element of a toast notification.* **(d) True**. *Toast notifications include a parameter value that is not displayed but is passed to the application if the user taps on the toast.*

For more information, see the section, "Notifying the Mobile Client of New Surveys," in Chapter 4, "Connecting with Services."

4. Why does the client need to register with MPNS before it can receive notifications?
   a. Because MPNS requires clients to authenticate before it will send notifications.
   b. Because MPNS can then notify your service that the client is ready to receive notifications.
   c. Because the client must obtain a unique URI to send to your service.
   d. Because the free version of MPNS has a limit on the number of clients who can receive notifications from your service.

   Answer: (a) False. *There's no requirement for MPNS clients to authenticate.* **(b) False**. *It is not the responsibility of MPNS to notify your service about clients. Clients must also register directly with your service.* **(c) True**. *Your client obtains a unique URI from the MPNS that it then forwards to your service. Your service can then pass the URI to the MPNS when it asks the MPNS to notify your client.* **(d) False**. *The free version of the MPNS has limits on the number of messages that you can send in a day, but does not limit the number of clients.*

For more information, see the section, "Notifying the Mobile Client of New Surveys," in Chapter 4, "Connecting with Services."

5. How does Tailspin transport data between the client and the web service?
   a. Tailspin uses the Microsoft Sync Framework to handle the data transport.
   b. Tailspin uses the Windows Communication Foundation (WCF) Data Service framework.
   c. Tailspin uses data transfer objects with a WCF REST endpoint.
   d. The mobile client application uploads directly to Windows Azure™ technology platform blob storage.

   Answer: (a) False. *However, Tailspin may consider this in the future.* **(b) False**. *However, Tailspin may consider this in the future when it fully supports Windows Azure table storage.* **(c) True**. *Tailspin currently uses this approach. The data transfer requirements are relatively simple, so this approach was not too difficult to implement.* **(d) False**. *This approach would not be secure or robust enough. The application also uses Windows Azure table storage.*

For more information, see the section, "Accessing Data in the Cloud," in Chapter 4, "Connecting with Services."

6. Why does Tailspin filter data on the server and not on the client?

a. To minimize the amount of data moved over the network.
b. To simplify the application.
c. For security reasons.
d. To minimize storage requirements on the phone.

Answer: (a) True. *Bandwidth costs are a significant consideration for any mobile application.* **(b) False**. *In some ways this complicates the application because the client has to send the filter criteria up to the server.* **(c) False**. *Not in Tailspin's scenario, but in other applications the filtering may determine what data the user is allowed to see rather than what they want to see.* **(d) True**. *This is not the primary reason for Tailspin but it does mean that the phone does not download data that is not relevant to the user and cache it while it filters.*

For more information, see the section, "Filtering Data," in Chapter 4, "Connecting with Services."

Index

Printed in Dunstable, United Kingdom

91343486R00118